Applied Geophysics with Case Studies on Environmental, Exploration and Engineering Geophysics

Edited by Ali Ismet Kanlı

Published in London, United Kingdom

IntechOpen

Supporting open minds since 2005

Applied Geophysics with Case Studies on Environmental, Exploration and Engineering Geophysics
http://dx.doi.org/10.5772/intechopen.78490
Edited by Ali Ismet Kanlı

Contributors
Yasir Bashir, Deva Prasad Ghosh, Román Alvarez, Gerardo Figueroa, Andrea Ustra, Vagner Elis, Olga Khachay, Oleg Khachay, Lawrence Hutchings, Seth Saltiel, Brian Bonner, Bing Zhou, Roberto Balia, Pier Paolo Manca

First published in London, United Kingdom, 2019 by IntechOpen
IntechOpen is the global imprint of INTECHOPEN LIMITED, registered in England and Wales, registration number: 11086078, The Shard, 25th floor, 32 London Bridge Street
London, SE19SG – United Kingdom
Printed in Croatia

British Library Cataloguing-in-Publication Data
A catalogue record for this book is available from the British Library

Additional hard and PDF copies can be obtained from orders@intechopen.com

Applied Geophysics with Case Studies on Environmental, Exploration and Engineering Geophysics
Edited by Ali Ismet Kanlı
p. cm.
Print ISBN 978-1-83880-740-5
Online ISBN 978-1-83880-741-2
eBook (PDF) ISBN 978-1-83880-742-9

For EU product safety concerns:
IN TECH d.o.o., Prolaz Marije Krucifikse Kozulić 3, 51000 Rijeka, Croatia, info@intechopen.com or visit our website at intechopen.com.

Meet the editor

Prof. Dr. Ali Ismet Kanlı received his undergraduate degree in 1989 at Istanbul University, Faculty of Engineering, Department of Geophysical Engineering. In 1994, he graduated from Istanbul University, Institute of Science, with an MSc, and in 1998 he completed his doctorate at the same institute. He started his academic career in 1992 as a research assistant in Istanbul University Faculty of Engineering, Department of Geophysical Engineering, Division of Applied Geophysics. He started to work as an assistant professor in 2001, associate professor in 2010, and professor in 2016 at the same division. Prof. Dr. Ali Ismet Kanlı is the head of the Applied Geophysics Division of the Geophysical Engineering Department. He has carried out and directed many international and national projects and has published several national and international scientific publications. He has been serving as a member of the editorial boards of 10 international journals and has been a reviewer of many international and national journals. He has also been a referee in many international and national projects. He is a member of three national and eight international scientific associations. Prof. Dr. Ali Ismet Kanlı's scientific interest areas are applied and near-surface geophysics, engineering and environmental geophysics, engineering seismology, exploration seismology, structural geophysics, earthquake engineering, geotechnical geophysics, borehole geophysics and well logging, alternative energy, and geothermal exploration.

Contents

Preface

Applied geophysics is based on physics principles that collect and interpret data on subsurface conditions for practical purposes, including oil and gas exploration, mineral prospecting, geothermal exploration, groundwater exploration, engineering applications, archeological interests, and environmental concerns. The depth of applied geophysics is shallow, typically from the ground surface to several kilometers deep where economic, cultural, engineering, or environmental concerns often arise. Applied geophysics uses almost all of the current geophysical methods, including electrical, magnetic, electromagnetic, gravimetric, geothermal, seismic, seismoelectric, magnetotelluric, nuclear, and radioactive methods. In applied geophysics, geophysicists are usually required to have a good understanding of math and physics principles, knowledge of geology and computer skills, and hand-on experience of electronic instruments. A geophysicist's routine job includes survey designs, data acquisition, data processing, and data interpretation with detailed explanation of the study.

This book provides a general introduction to the most important methods of applied geophysics with a variety of case studies. These methods represent a primary tool for investigation of the subsurface and are applicable to a very wide range of problems. Contributors to this book have made an effort to put their ideas into simple terms without forgoing quality. The text covers the state of the art and future directions in relation to applied geophysics.

The principal object of this book is to place all the recent interesting and technical literature in one location, particularly the analysis of seismic responses using non-linear methods, radiometric mapping of hydrothermal alterations, wave modeling and subsurface seismic imaging, geotechnical modeling with seismic tomography, electrical resistivity tomography, fine-scale depositional system studies with high-resolution sequence framework, induced polarization/resistivity applications, and rock physics interpretation in geothermal reservoirs.

I consider this book to be convenient for self-study by engineers and scientists in any discipline related to earth sciences, especially in applied geophysical methods.

Ali Ismet Kanlı, PhD
Professor of Geophysics,
Istanbul University-Cerrahpasa,
Faculty of Engineering, Department of Geophysical Engineering,
Avcilar Campus, Istanbul-Turkey

Chapter 1

Advance Wave Modeling and Diffractions for High-Resolution Subsurface Seismic Imaging

Yasir Bashir and Deva Prasad Ghosh

Abstract

Seismic modeling and Imaging for the small-scale feature in a complex subsurface geology such as salt deposit, fracture reservoir, and Carbonate is not casual because of propagated wave affected by many objects once it hits the geologic structure in the subsurface. The principal goal of newly developed seismic modeling & imaging is to get a subsurface image of structural features with greatest sharpness or resolution. Using model dataset the Sigsbee and Marmousi, we illustrate the accuracy of conventional and advance wave modeling techniques. However, in conventional a Finite difference (FD) algorithm is used to generate the data and in advanced wave modeling, the low-rank (LR) approximation is used to acquire zero-offset configuration data. A field dataset from Malaysian basin is re-processed and imaged using diffraction imaging which shows an enhancement in structural interpretation. Furthermore, the results gained from the proposed modeling and imaging approach significantly enhance the bandwidth of the imaged data. Finally, a frequency spectrum shows a recovery of low-frequency from 0 to 60 Hz which is an optimal resolution of seismic imaging.

Keywords: seismic modeling, finite difference, low-rank, diffraction imaging, plane-wave destruction, exploding reflector migration

1. Introduction

1.1 Malaysian basin

Malaysian basins are structurally complicated because their assessment through different phases of continental accretion, rifting and mountain building. The geology of the Malay Basin is very old and different that the other part of Sarawak and Sabah basin [1]. Geophysical and geological challenges include, fine sand imagery, often beyond seismic resolution; imaging under gas chimney and under carbonates; diffraction imaging; imaging of the internal architecture of the basement; understand the propagation of waves in efficient media and the associated anisotropy; speed analysis and anisotropy; Gas cloud imaging using complete waveform inversion; and multiple elimination [2]. The focus of research in this part of the world is to enhance the image quality of the faults, fracture and karst using diffraction imaging. **Figure 1a** shows the geographical location of the area which is near to the Borneo main land and a part of Sarawak Basin. **Figure 1b** is the cross section taken

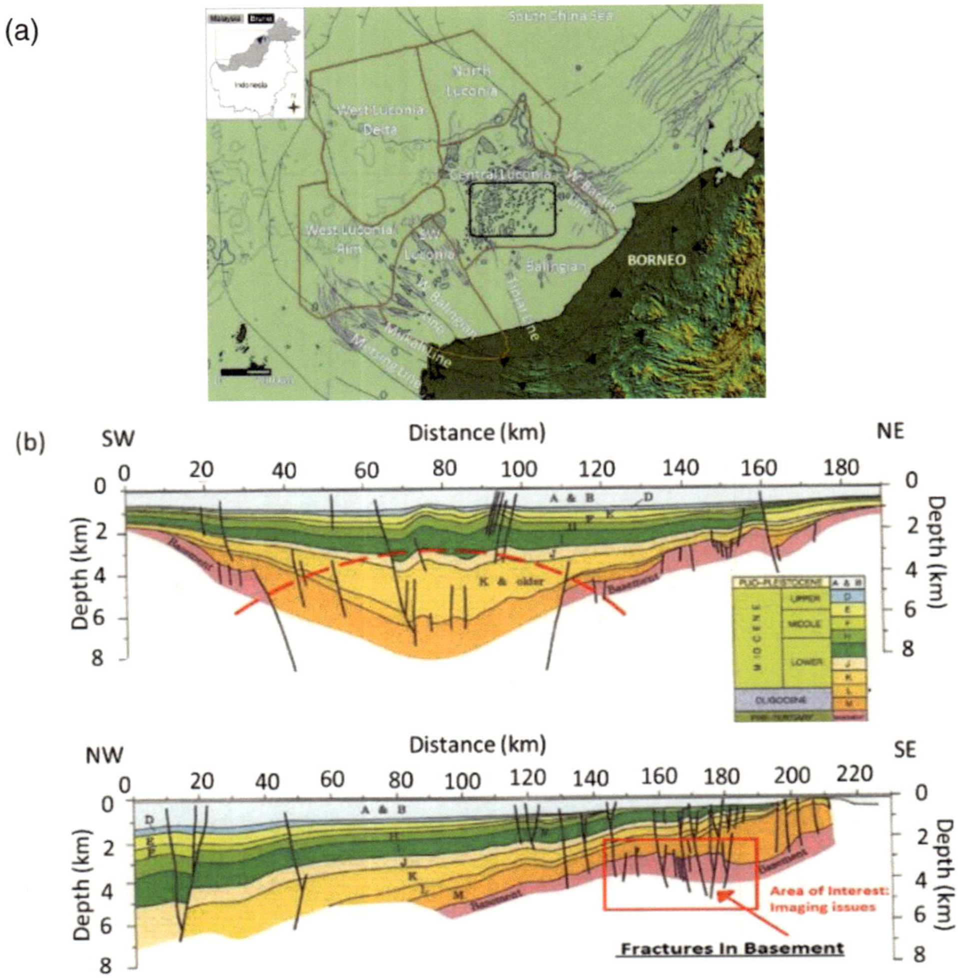

Figure 1.
(a) Location map of the Sarawak Basin [3] and (b) a cross section of Malay Basin subsurface structure. Fractured basement located at a depth of 2.0–5.0 km, varying with lateral extension [2].

from Malay Basin subsurface geometry which is an extensional regime and because of the tectonic activity the normal faulting can be seen.

1.2 Sarawak basin

Diffraction imaging analyzes were carried out in the Sarawak Basin, located north-west of Borneo, forming the southern boundary of the Oligocene-Recent Basin of the South China Sea; its tectonic evolution has been almost matched by seabed rifting and spread in the marginal basin of the South China Sea [4]. The majority of the Sarawak Basin, including this part of Malaysia, is composed of carbonates, making seismic imagery difficult. The pre-carbonated Oligocene deposit in the lower Miocene and later the terrestrial deposit filled the shelter zone in Cycle I and Cycle II. The cycles are described as follows: Cycles I and II (Upper Eocene to Lower Miocene) have been interpreted as channel sands, lowered clays, and deposited coal. Cycle III (lower-middle Miocene) contains fine limestone shale and sandstone, cycle IV (middle Miocene) is composed of limestone with few mixed clastics, cycle V (middle to upper Miocene) is recognized as limestone, cycle

VI to cycle VIII (upper Miocene to Pleistocene) is composed of open marine and coastal clays and sand, respectively. The progradante sediments from cycle V1 to cycle VIII have gradually stifled the large accumulations of carbonates to the present day. In the center of Luconia, carbonate deposits began at the beginning of the Miocene (cycle III) and increased significantly in the middle and late Miocene cycles IV and V [3]. The structure is dominated by a simple extension phase NS trend faults. These faults affected the deposits during cycles I and II and served as foci for the development of carbonate reefs of cycle IV and V and their accumulation. The dominance of intensive pre-carbonation structuring has therefore resulted in poor seismic image quality.

1.3 Diffraction

Diffraction hyperbolic patterns occur frequently in recorded seismic data, particularly in carbonate reservoirs due to abrupt lateral changes in impedance contrast and discontinuity of subterranean layers. However, a very serious doubt to the application of classical theory has claimed that stacked seismic data is not true zero separation data because it is not clear that the results of the stacking of data recorded on a wide range of source-receiver separations is practically close to the results of real zero separation recording, since diffraction amplitudes are concerned [5]. Berryhill also explained the concept and compared the theory of zero separation with source-geophone distance without zero separation, concluding that the diffraction amplitudes at a source-receiver separation different from zero and well known with respect to the wave and geometric propagation paths, a point diffractor gives rise to a hyperbolic diagram on the stacked section. Hyperbola is explained as a symmetrical open curve formed by the intersection of a circular cone with a plane at a smaller angle with its axis than the sides of the cone. Diffraction only be considered as hyperbolic if the recovery is homogeneous, which not a very natural supposition to make is. The curvature of the diffraction hyperbola depends on the speed of the medium, the apex being an indicator of the location of the defect. Diffraction imaging is a dare in seismic processing and adopted by a workflow by means of the common reflection surface by [6–8].

In reflection seismology applied to exploration geophysics, such wave propagation phenomena are employed to estimate the properties of the Earth's subsurface reflector. Furthermore, the diffraction phenomenon is also concerned with reflection because of the properties of the subsurface as defined above. The acoustic (seismic) impedance, Z=, Where V is the seismic wave velocity and ρ is density. Although seismic migration is now the one of the primary imaging tools employed in the field, the earliest analogue seismic records took the form of simple single-fold illustration [4, 9–12]. These recordings were characterized by diffracted energy and random noise, but they nonetheless provided a useful interpretation of the subsoil of the Earth. Later, mechanical migration eliminated the structural deformation of early seismic data, with CMP stacks condensing the amount of random noise as the diffracted energy is preserved. Seismic reflection and diffraction waves are essentially different physical phenomena. Most seismic processors tune and image by improving the seismic reflection data, and do not deliberate the diffracted waves present in the processed data, which carry most of the information regarding minor but important subsurface events [13]. Such small-scale underground events (such as faults, fractures, channels, karsts, and salty body edges) occur as diffracted waves in the seismic data, which can be captured [14, 15].

The plane wave destruction filter (PWD) was originally introduced by Claerbout [28] for the characterization of seismic images using the superposition of local plane waves. This PWD filter was based on the plane wave differential equation, after the original plane wave destruction filter with the same approximation proved poor when applied to spatial folding data [16]. On the other hand, the dip frequency filtering approach is applied in the f-k domain. Here we use the Fourier transform to convert time data to the frequency domain, with a filter designed to eliminate reflections based on wave cycles per kilometer. In this paper, we develop a workflow that captures these events on a small scale through separate diffractions based on the regularity and continuity of the slope of the local event that corresponds to the reflection event. We compare the two techniques of dip-frequency filtering and wave-destruction filtering, before integrating the two approaches and performing comparative analyzes on the optimal preservation of diffractions.

1.4 Diffraction imaging and resolution

Seismic resolution and Diffraction imaging have a direct relationship to each other, while the frequency of the seismic wave also affects seismic image, a high frequency providing a high resolution image. The relationship between seismic wave frequency and resolution, wavelength, penetration, and hyperbolic diffraction curve is shown in **Table 1** [17, 18].

The resolution of seismic data is typically around λ/4, which means that high frequency waves illuminate a small object; however, as the increase in frequency will also affect penetration depth, it is usually defined by the purpose of the research. Seismic surveys typically use lower frequencies for subsurface imaging. As a result, seismic images have a slightly lower resolution (λ/4 to λ/8 dependent on data quality) but a deeper penetration (10 km) than those produced by biomedical imaging, requiring a lower depth of penetration (in feet) but higher resolution on the acquired frequency of the data.

1.5 Diffraction imaging in depth and time

Although many studies have been conducted on diffractions based on both depth and time domains, the latter does not provide an accurate indication of the discontinuity of the subsurface feature examined in this paper. In the field of depth, the existence of diffraction is a frequent indicator of high complexity and a strongly inhomogeneous trend [19]. As a result, pre-stacking depth imaging methods require more work and computation than those using time imaging [20]. In addition, the success of diffraction identification and isolation in the depth domain depends on the accuracy of the velocity model used to provide the full wave depth image.

1.6 Multi-focusing diffraction imaging (MFDI)

MFDI is a novel temporal displacement correction method that effectively describes diffraction events, with the ideal sum of diffracted events and the

Higher frequency	Lower frequency
Low depth of penetration	High depth of penetration
High resolution	Low resolution
Small wavelength	Large wavelength
Lower diffraction response	Higher diffraction response

Table 1.
The relationship between seismic wave frequency and other parameters.

attenuation of specular reflections allowing the creation of an image mainly comprising diffraction energy. The time correction, which is based on the multiple focus method, depends on two parameters, the emergence angle and the radius of curvature of the diffracted wavefront. The above parameters are calculated from the seismic traces of the pre-stack [19]. The result of the IFM is therefore a high-resolution full-azimuth seismic image that comprises optimally stacked diffraction events. The diffraction section contains important data on local heterogeneities and discontinuities in the geology of the subsoil, which can be used to improve horizontal drilling and determine the optimal location of exploration wells.

The multi-focus diffraction imaging makes it possible to:

- Identify naturally fractured reservoirs
- Avoid unwanted liquid pathways
- Cartographic sources of anisotropy of speed
- Contour defects and salt body

1.7 Finite difference modeling

Finite difference methods (FDM) are extensively used in seismic modeling and migration. In this chapter, a conventional FDM model was used for modeling, with model input being velocity and density values, and the model produces seismic data. FDM are numerical methods for solving differential equations by approximating them with differential equations, in which finite differences approach derivatives. In seismic wave modeling, FD methods are used to propagate the wave in the subsoil. This method has no immersion limitation and produces all events associated with the wave equation such as multiple reflection, head waves and elastic wave equation, anisotropic effects, and wave conversion data [21]. Therefore, the modeling of the F-D wave equations is the ideal way to produce the synthetic seismic data. However, the ultimate goal of migration is to get the image of the real earth using seismic data, which is difficult to test the accuracy of the migration methods with the desired results. In case of seismic inversion, the entry includes the traces and the output of the structural image. For this purpose, the quantities below must be calculated [22]:

$\varnothing$ wave propagation angle
τ two way travel time
β angle of incident from source or receiver
$\partial\beta/\partial x$ geometrical spreading parameter

For each source or receiver, the above quantities must satisfy the equations [22].

$$\left(\frac{\partial\tau}{\partial x}\right)^2+\left(\frac{d\tau}{dx}\right)^2=\frac{1}{v^2(x,z)} \tag{1}$$

$$\sin\varnothing=v\frac{d\tau}{dx} \tag{2}$$

$$\frac{\partial\sigma}{\partial x}\frac{\partial\tau}{\partial x}+\frac{\partial\sigma}{\partial z}\frac{\partial\tau}{\partial z}=1 \tag{3}$$

$$\frac{\partial\beta}{\partial x}\frac{\partial\tau}{\partial x}+\frac{\partial\beta}{\partial z}\frac{\partial\tau}{\partial z}=1 \tag{4}$$

$$\frac{\partial}{\partial z}\left(\frac{\partial \beta}{\partial x}\right)+\frac{\partial}{\partial x}\left[\mu(x,z)\frac{\partial \beta}{\partial x}\right]=0 \quad (5)$$

where v(x, z) is the velocity and $\mu(x,z)=\frac{\partial \tau}{\partial x}\left[\frac{\partial \tau}{\partial z}\right]^{-1}$.

Eq. (1) is the Eikonal equation, Eqs. (3) and (4) are derived from Pusey and Vidale [23] and Eq. (5) follows from Eq. (3).

1.8 Plane-wave destructors (PWD)

Plane wave destruction filter is derived from the local plane wave model for characterizing the seismic data. This filter operates in the time domain (T-X), such as the time distance, and can be extended to the frequency domain. PWD is constructed using an imbedded finite difference scheme for the local plane wave equation as described [24].

For the characterization of several plane waves, it is possible to flow several filters similar to that of the following equation:

$$A(Z_x)=\left(1-\frac{Z_x}{Z_1}\right)\left(1-\frac{Z_x}{Z_2}\right)\ldots\left(1-\frac{Z_x}{Z_N}\right) \quad (6)$$

where Z1, Z2, ..., ZN are the zeros of the polynomial. The Taylor series method (assimilating the coefficients of the expansion of the Taylor series to the zero frequency) gives the expression

$$B_3(Z_t)=\frac{(1-\sigma)(2-\sigma)}{12}Z_t^{-1}+\frac{(2+\sigma)(2-\sigma)}{6}+\frac{(1+\sigma)(2+\sigma)}{12}Z_t \quad (7)$$

For a three-point centered filter B3 (Zt)

$$\begin{aligned}B_5(Z_t)=&\frac{(1-\sigma)(2-\sigma)(3-\sigma)(4-\sigma)}{1680}Z_t^{-2}+\frac{(4-\sigma)(2-\sigma)(3-\sigma)(4+\sigma)}{420}Z_t^{-1}\\&+\frac{(4-\sigma)(3-\sigma)(3+\sigma)(4+\sigma)}{280}+\frac{(4-\sigma)(2+\sigma)(3+\sigma)(4+\sigma)}{420}Z_t\\&+\frac{(1+\sigma)(2+\sigma)(3+\sigma)(4+\sigma)}{1680}Z_t^2\end{aligned} \quad (8)$$

For a five-point centered filter B5 (Zt), the derivation of Eqs. (7) and (8) can be found in Fomel [24].

The filter used in the present work is a modified version of filter A (Zt, Zx):

$$C(Z_t,Z_x)=A\ (Z_t,Z_x)B\left(\frac{1}{Z_t}\right)=B\left(\frac{1}{Z_t}\right)-Z_xB(Z_t) \quad (9)$$

This filter avoids the requirement of a polynomial partition. In the case of the three-point filter Eq. (7) and the two-dimensional filter Eq. (7), there are six coefficients consisting of two columns; in each column there are three coefficients and the second column is an inverse copy of the first one. However, the decomposition algorithm is significantly more expensive than the FD. This algorithm is extended to Song's anisotropic wave propagation in 2013 by involving the Eigen function rather than rows and columns of the original extrapolation matrix [25] (**Figure 2**).

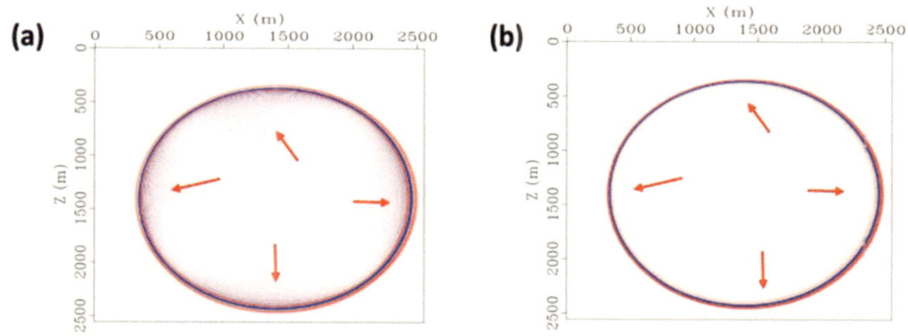

Figure 2.
(a) Snapshot of a wave field in a smooth velocity model calculated using the fourth order finite difference method; and (b) the lower rank approximation of the wave field in the same smooth velocity model [26].

1.9 Slope estimation

Slope estimation is a necessary step in applying the FD plane-wave filters to real data [27], although estimating dissimilar dual slopes, σ_1 and σ_2, in the available data is complicated than estimating a single slope [24].

The regularization condition should thus be applied to both $\Delta\sigma_1$ and $\Delta\sigma_2$ as follows:

$$\varepsilon D \Delta\sigma_1 \approx 0 \tag{10}$$

$$\varepsilon D \Delta\sigma_2 \approx 0 \tag{11}$$

The solutions of the above equations depend on the primary values of slop. 1 and slop. 2, which should not be equal, but can be prolonged to the numerical equation with respect to the grid number of the dataset. However, this equation is used here to calculate slopes for the given data set. In the current study, we used a reformed and better version of the plane wave destruction method for seismic diffraction parting based on Claerbout [28].

1.10 Diffraction separation methods from seismic full-wave data

One of the best practices and methods of diffraction preservation is the plane wave destruction (PWD) filter initially introduced by Claerbout [28] for the description of seismic images using local plane wave superposition. This PWD filter is based on the plane-wave differential equation, after the original wave-destroying filter plane with the same approximation showed poor performance when applied to spatially aliased data in comparison with other filters. Frequency-distance prediction error [16]. Planar Wave Destruction Filter, which can be considered as a time-distance (TX) analog of the frequency-distance (FX) prediction error filter, derived from a local plane wave model is used to characterize the seismic data [24]. Unfortunately, the first experiments in applying flat-wave destruction for spatially-aliased data interpolation [28] have shown poor results compared to standard frequency distance prediction error (FX) filters [16]. A workflow that uses plane wave destruction for diffraction imaging is shown below. Flat-wave destruction of common shift data may have difficulty in extracting diffractions in regions with complex geological and velocity variations [11] (**Figure 3**).

2. 2D examples of a complex faulted model: Marmousi

The proposed methodology is tested on a model called Marmousi, created in 1988 by the French Petroleum Institute (IFP) [30]. This model contains 158 horizons with horizontal layers and series of normal faults, which makes it complex especially at the center of the model. This model length is 9.2 km and the depth is 3 km. **Figure 4(a)** shows the Marmousi model, in which we assume that our task is to imagine the structure of the anticline below and consider it as a reservoir. We use a Ricker wavelet at a point source with a dominant frequency of 40 Hz. The size of the horizontal grid x is 4 m and z is 4 m. **Figure 4(b)** shows the result of the results of the advanced wave modeling using a low rank approximation. This modeling technique is capable of carrying out different designed studies, but for this we have selected the zero shift data record.

This imaging workflow is a new development for high-resolution imaging, as shown in **Figure 3**. We tested the workflow by separating the diffraction and the residue and representing it separately. **Figure 5**(a) shows the diffraction separated using plane wave destruction filtering and **Figure 5**(b) is the residue after diffraction separation. **Figure 6(a)** shows the zero shift data migrated with full wave imaging and (b) is the imaged section combining both reflection migration and diffraction migration. This shows an improvement in the resolution especially in the amplitude of faults and small discontinuities that are not resolved by the conventional imaging method.

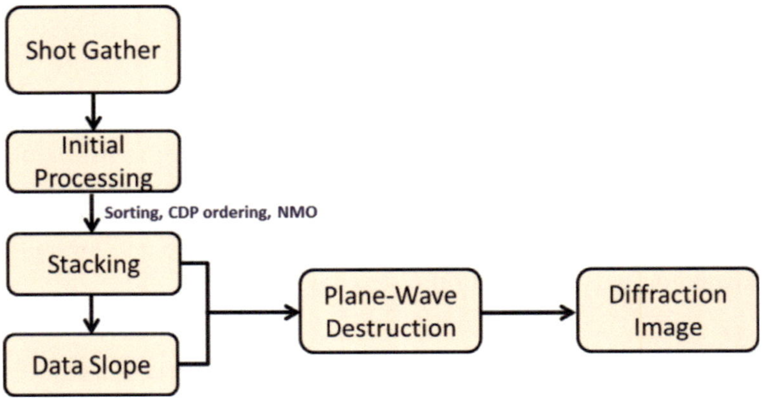

Figure 3.
A generalized diffraction separation workflow using plane wave destruction (PWD) based on estimating the slope of the data before imaging [29].

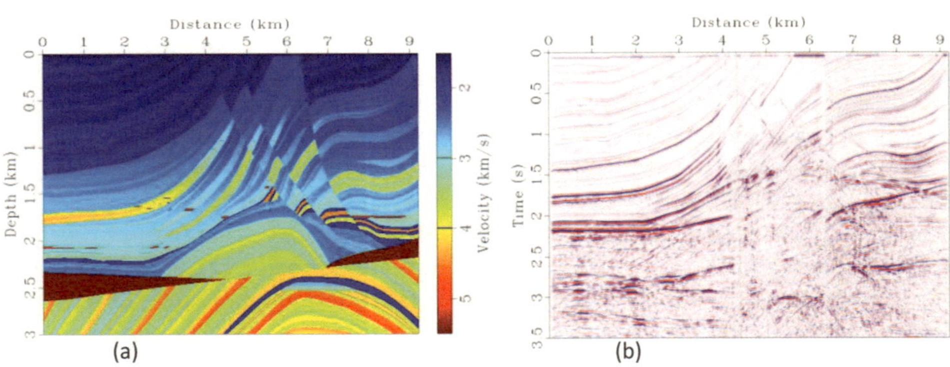

Figure 4.
(a) Marmousi velocity model and (b) modeled seismic data using low-rank approximation.

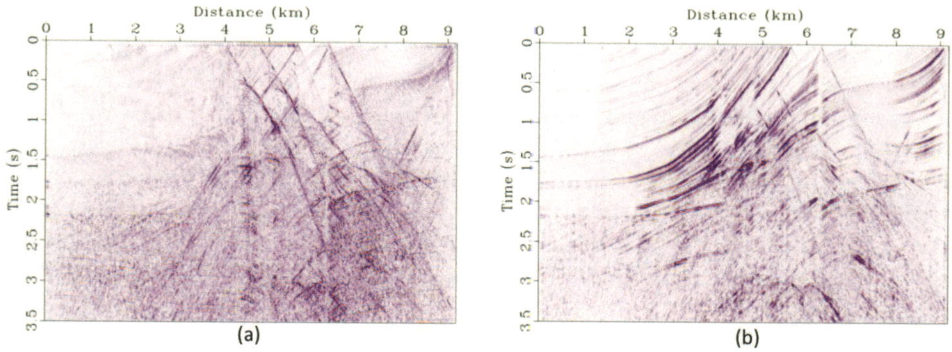

Figure 5.
Separation of the seismic data (a) separated diffraction and (b) separated reflection.

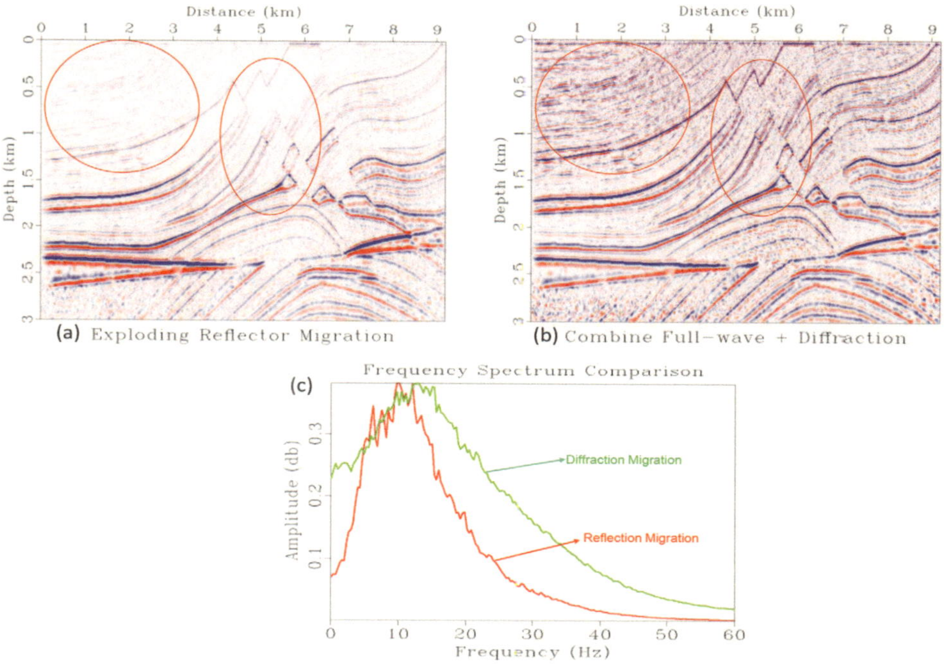

Figure 6.
Imaging of the seismic data (a) conventional zero-offset migration, (b) reflection and diffraction migration combined and (c) frequency spectrum of the data a & b [31].

For quantitative interpretation of the results, **Figure 6**(**c**) shows a frequency spectrum for both conventional (red) and diffractive (green) imaging. Higher diffraction response. In addition, higher frequency data of 50–60 Hz is enhanced for high resolution imaging.

3. Application of method in field data: Sigsbee (Gulf of Mexico)

Most information obtained from high resolution or super-resolution images is the result of careful processing. Consideration of diffraction is necessary in advanced wave modeling and processing because in conventional processing, diffractions are suppressed intentionally or implicitly [32]. In addition, the separation of these diffractions using plane wave destruction filtering could preserve the diffracted amplitude. To prove the proposed modeling and diffraction migration results, we examined the Sigsbee 2A model to study wave extrapolation in a complex velocity field model containing a sedimentary sequence fragmented by a number of normal faults and overlapping (**Figure 7a**). In addition, there is a complex salt structure in the model that causes illumination problems as part of the current processing and imaging approach. These imaging problems allowed us to develop appropriate algorithms for better imaging. The Sigsbee 2A model has an absorbent free surface condition and a lower than normal water body reflection. These properties do not generate the effect of free surface multiples and lower than normal internal multiples.

For seismic imaging purposes, a zero shift design was chosen to acquire the zero offset seismic data for simplicity, as shown in **Figure 7b**. Advanced modeling of low-rank waves is used to obtain non-dispersive seismic. The comparison of two waves modeling the finite difference and the lower row is performed for high resolution imaging purposes. **Figure 5** shows an evaluation of the results of conventional **Figure 8(a)** and advanced **Figure 8(b)** wave modeling and imaging results that indicate that saline body edges are unresolved. Modeling and imaging.

The plane wave destruction method is used to separate the diffractions from the full-wave reflection data. It is important to rely on the estimated dip of the data and to recognize an accurate determination of the sinking wave as it is an essential parameter for plane wave destruction filtering to separate diffraction and reflection. In this research, we separated the diffractions from the complete wave and tried to imagine these diffractions with a correct velocity model (**Figure 9b**). **Figure 9a** is the result of a depth-migrated image using conventional wave modeling results, which are poor compared to the results of advanced wave modeling and diffraction

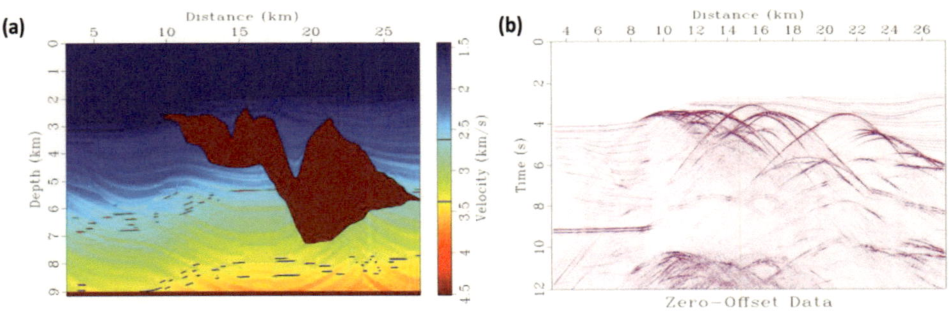

Figure 7.
(a) Sigsbee velocity model, from Gulf of Mexico (GOM) (b) zero-offset seismic data using low-rank approximation.

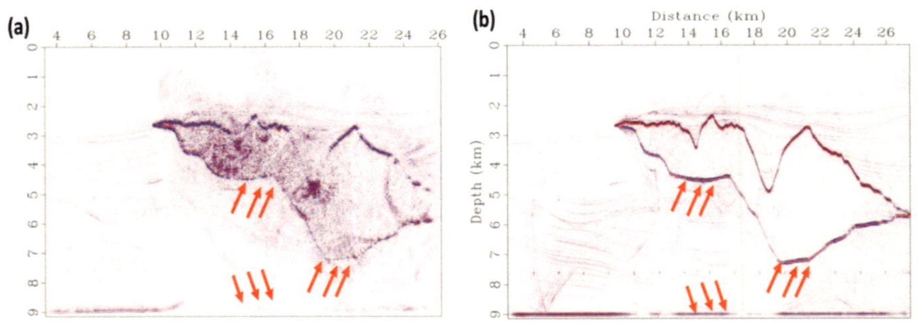

Figure 8.
Seismic imaging of full wave data using (a) conventional finite difference modeling and (b) advance wave modeling using low rank approximation.

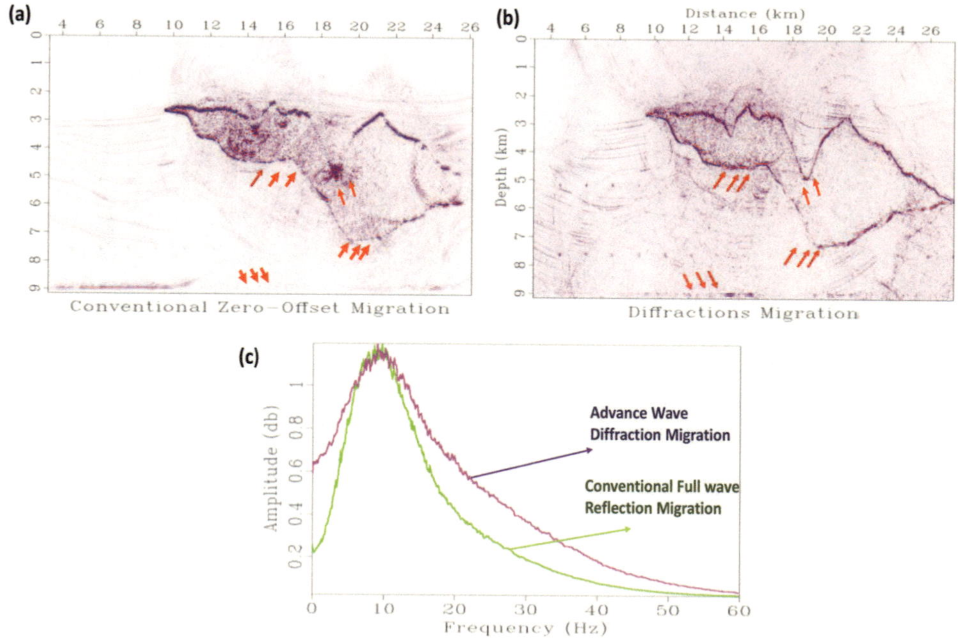

Figure 9.
Reflection and diffraction migration (a) conventional zero-offset migration, (b) diffraction migration and (c) frequency spectrum comparison of conventional full wave migration and advance wave approximation diffraction migration.

imaging. **Figure 9c** shows the frequency spectra of the migrated data. The original offset using classical modeling and full wave migration with Fourier migration in several steps is indicated in green, and advanced waveform modeling with diffraction migration is displayed in purple. Improvements in diffraction imaging, especially for small scale events, are highlighted in the spectrum, allowing us to achieve high resolution images with low frequency data recovery after migration. Low frequency signals from long-lived seismic data are critical values for many areas of exploration seismology and hydrocarbon prediction.

4. Application of method in field data: Malaysian basin

The angular stack is calculated to measure the reflectivity of a particular angle, the angle stacked data are used to observe the amplitude over offset (AVO) for the Direct Hydrocarbon Indicators (DHI) and the inversion in the oil and gas industry [33, 34]. The stacking of angles also applies to the overall combination of interceptions and gradients. These angle stacks typically have a near, mid, and far angle, but an angle stack may have more than three angle stacks with a minimum of 1°. Here, for the diffraction studies, a stack of 3 angles has been realized, as shown in **Figure 10**. For the comparison on the near and far, the stacking is performed with two angles of 4.5 and 31.5 degrees, as shown in **Figure 11**. A near angle provides the best diffraction amplitude rather than a far angle stack in this study area.

As the stacking angle rises, there is a loss of amplitude in the data, as can be seen in the lower section (**Figure 10**). A point diffraction that can be observed in the data (2150 ms) shows the behavior of the diffraction change; the diffraction flanks are not wide in the stack of nearly staggered angles, while in the offsets, the flanks of the curves are wider. This diffraction would require a greater migration aperture to stack the energy of the top diffraction flanks (**Figure 11**).

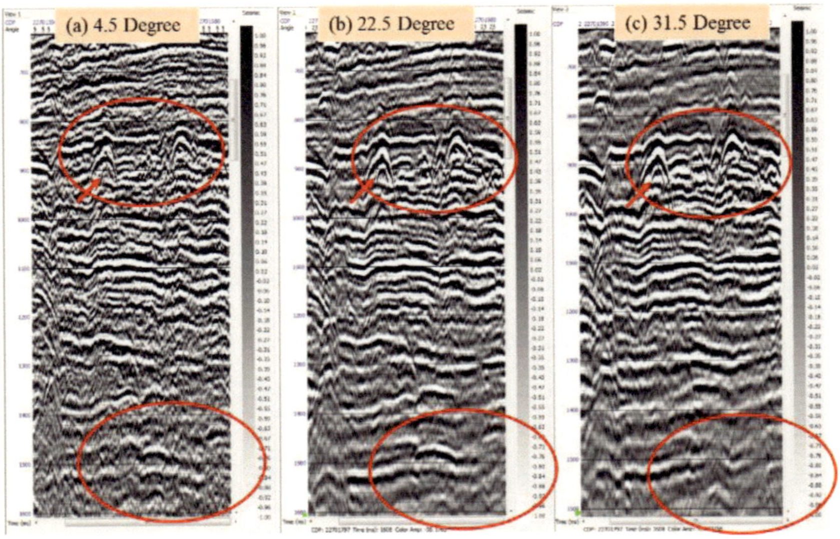

Figure 10.
Angle stack of three different range (a) range limited stack data with 4.5°, (b) 22.5 degrees and (c) 31.5°.

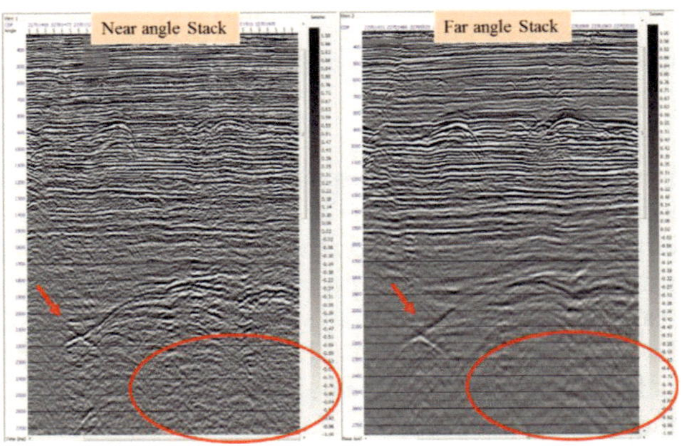

Figure 11.
Partial stack seismic: (a) near angle stack 4.5°, and (b) far angle stack 31.5°.

4.1 Influence of offset on diffraction imaging

Figure 12 illustrates the diffraction with different range of offset. Theoretically, the sides of the diffraction curve are influenced by the velocity and time/depth of the point diffractor [35]. However, **Figure 13** shows that the diffraction also depends on the data shift because the near offset data has less diffraction response than the far-lag data. The reason for this is that the far-lag data covers more offsets and therefore recorded the wider angle data set.

4.2 Methodology for diffraction separation for high-resolution imaging: Real field data

Non-migrated offset collection data was on condition that for this project. The initial processing, including the sorting of the common midpoint offset (CMP) collection, was performed to obtain the seismic section of the stack. The following processes were implemented in the sorting process:

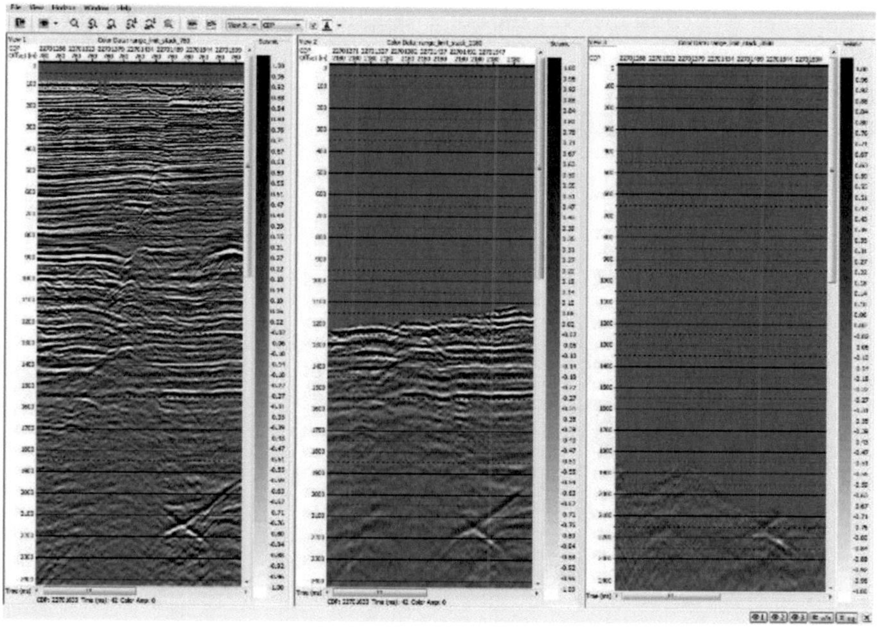

Figure 12.
Multiple offsets gather data shows that shallow data is not recorded in the far offset (a) range limited stack of 0–780 m (b) 780–2180 m and (c) 2180–3580 m.

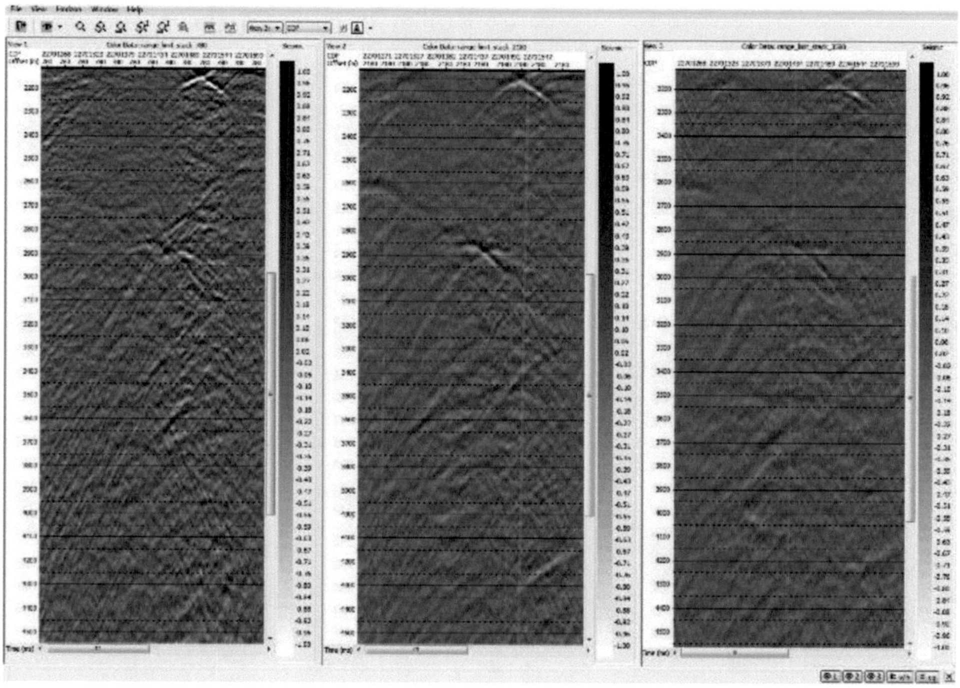

Figure 13.
Amplitude decay with offset is recorded (a) 0–780 m, (b) 780–2180 m offsets and (c) 2180–3580 m offset stack, signal-noise ratio and amplitude is stronger in the near offset.

1. Window selection around the desired structure with the maximum diffraction response.

2. Select the inline or crossline 3D data for 2D analysis.

3. The inline was constant over the entire length; 810 traces were extracted from the transverse line.

4. Perform an NMO and velocity analysis.

5. Perform offset-dependent diffraction enhancement.

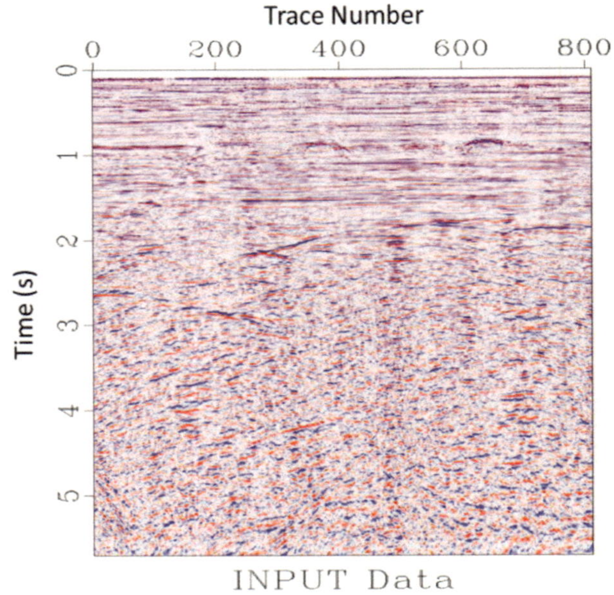

Figure 14.
Actual input data of the Malaysian basin. Treated with careful diffraction treatment and stacked data prior to migration: Sarawak Basin, carbonate accumulation structure.

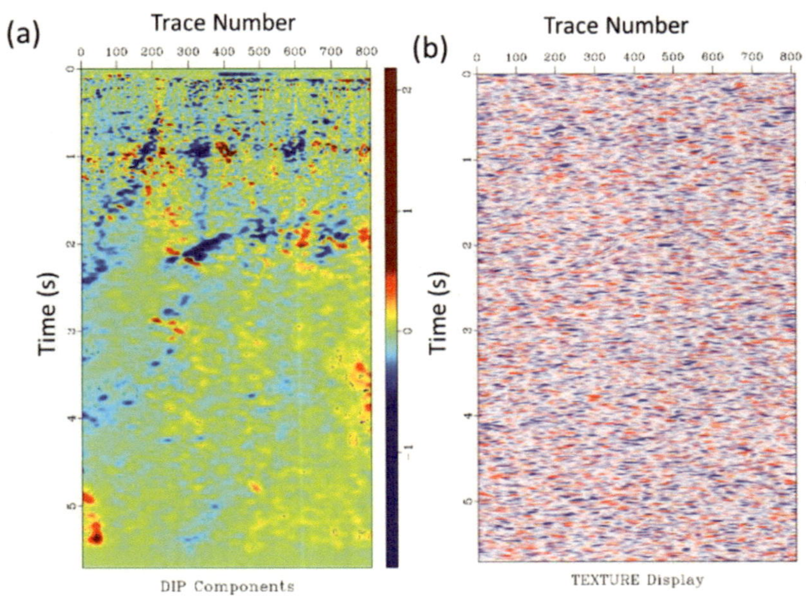

Figure 15.
(a) Predicted dip field of data shown in ***Figure 14****, and (b) texture computed by convolving field number with the inverse of plane-wave destruction filters.*

6. Stack the data for diffraction analysis in the full stack data set.

7. Estimate the dip components from the data using Eqs. (10) and (11) given above.

8. Remove reflections and preserve diffractions via PWD filtering (Plane-wave Destruction) [24].

Figure 14 shows a 2D line extracted which is an unmigrated seismic from a carbonated field in the Sarawak Basin. It has been carefully treated using pre-imaging procedures. The diffraction separation method was then extended to preserve the diffractions in the actual data. **Figure 15a** shows the estimated dip components of the data, which help identify tilt faults and pinches, while **Figure 15b** shows the corresponding texture [36] obtained by convolving a random number field with the inverse of The destruction by plane wave filters The latter was built

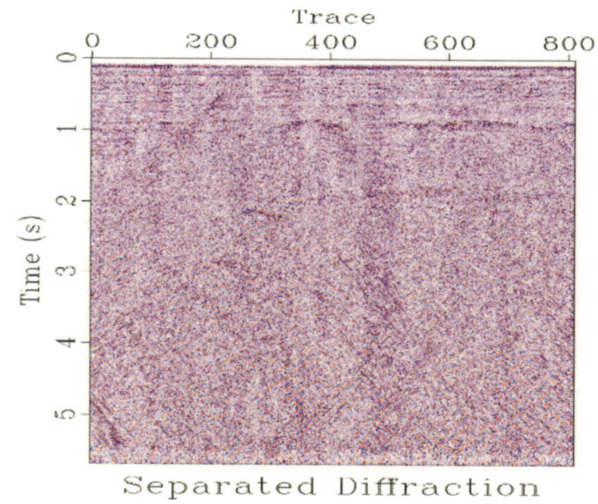

Figure 16.
Unspoiled diffractions after the application PWD filtering on stack data.

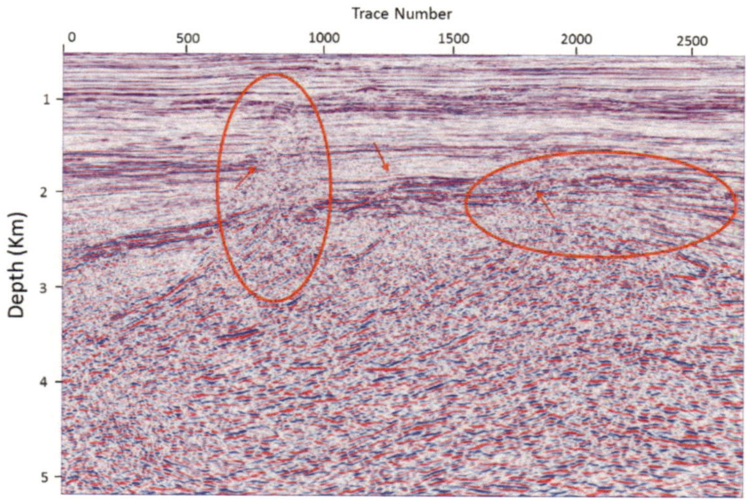

Figure 17.
Imaging section, including preserved diffraction. The fractures are resolved and the quality of the seismic data is improved.

using helical filtering techniques [37, 38]. The advantage of displaying the texture is to visualize the characteristics of the local plane in the data with the dip. **Figure 16** shows the separate diffraction which is the input for diffraction imaging. This diffraction data is migrated separately and merged with the residual data migration shown in **Figure 17**. The final result of the migration including the diffraction and reflection data which improved the resolution of the data. Inside the red circle on the left side, a major fault is imaged and can be interpreted, further on the right side in the red circle small scale faults are illuminated after imaging.

5. Conclusion

We present the significance of advance wave modeling and significance of careful pre-image processing and diffraction imaging in the multifaceted earth like fractured zones, fault edges, small-scale faults and pinch-outs. The effect of the Angles and offsets also be incorporated in the study by concluding that a far stack data has a higher diffraction response than the near stack seismic data. The analysis of different offset gather data shows the observation of the diffraction with the angle of an offset. The results of the far offset data provide the higher diffraction response because of the coverage of the long offset in the data.

Plane-wave destruction filters with an improved finite-difference design have added value in processing for maintaining diffraction. These diffraction data contains the information of the diffracted events which are not recorded in the reflected wave, and further migrate it separately and merge with the reflected data. Finally, considering diffraction in the processing and imaging is very useful for the high-resolution diffraction imaging.

Acknowledgements

The authors are thankful to Universiti Teknologi PETRONAS (UTP), the Geoscience Department and the Centre of Seismic Imaging (CSI) for providing the facilities for this research work. We would also like to thank PETRONAS for funding this work and providing data for research and publication.

Conflict of interest

The authors declare no conflict of interest.

Notes/Thanks/Other declarations

It is a great pleasure to thank those who made this research possible, such as my ever supporting parents, family members who gave me the moral support I required. I am heartily thankful to my co-author, Prof Dr. Deva Prasad Ghosh, whose encouragement, supervision and support from the preliminary to the concluding level enabled me to develop an understanding of the subject. I would like to make special thanks to my co-supervisor A.P Dr. Chow Weng Sum for supervising, guiding and moral support to complete my research. A special thanks goes to my department collogue which always welcome to review my research work and kind suggestions to improve the research and articles, those include but not limited to Prof. Abdul Ghani, Dr. Ahmed Salim, A.P Lo Shyh-Zung, Dr. Hassan, Dr. Suhaili,

A.P Wan Ismail, Khairul Ariffin, Abdul Halim, Siti Nur Fathiyah, Luluan Lubis. My special appreciation goes to Prof. Sergey Fomel and Luke Decker from The University of Texas Austin & Texas Consortium for Computational Seismology for their valuable advice and guidance over the course of this study.

I would also like to thank my colleagues in Center for Seismic Imaging (CSI) who assisted and supported me in the completion of this project including Dr. Yaser, Dr. Sajid, Dr. Iftikhar, Dr. Maman, Amir, Annur, Liu, Teresa, Hammad, and Irfan and to all CSI members. Many thanks for discussion and sharing of knowledge during the research work.

Author details

Yasir Bashir* and Deva Prasad Ghosh
Centre of Excellence in Subsurface Seismic Imaging and Hydrocarbon Prediction (CSI), Geosciences Department, Universiti Teknologi Petronas (UTP), Seri Iskandar, Perak, Malaysia

*Address all correspondence to: dryasir.bashir@live.com

References

[1] Madon M. Geological setting of Sarawak. In: The Petroleum Geology and Resources. Malaysia; 1999. pp. 273-290

[2] Ghosh D, Abdul Halim MF, Viratno B, Darman N. Geophysical issues and challenges in Malay and adjacent basins from and E & P Perspective. The Leading Edge. 2010;**4**:436-449

[3] Janjuhah HT, Gamez Vintaned JA, Salim AMA, Faye I, Shah MM, Ghosh DP. Microfacies and depositional environments of Miocene isolated carbonate platforms from central Luconia, offshore Sarawak, Malaysia. Acta Geologica Sinica - English Edition. 2017;**91**(5):1778-1796

[4] Madon M, Kim CL, Wong R. The structure and stratigraphy of Deepwater Sarawak, Malaysia: Implications for tectonic evolution. Journal of Asian Earth Sciences. 2013; **76**:312-333

[5] Berryhill JR. Diffraction response for nonzero separation of source and receiver. Geophysics. 1977;**42**(6): 1158-1176

[6] Dell S, Gajewski D. Common-reflection-surface-based workflow for diffraction imaging. Geophysics. 2011; **76**(5):S187-S195

[7] Krey T. The significance of diffraction in the investigation of faults. Geophysics. 1952;**17**(4):843-858

[8] Schwarz B, Gajewski D. Accessing the diffracted wavefield by coherent subtraction. Geophysical Journal International. 2017;**211**(1):45-49

[9] Fourier J. Sine and Cosine Series for an Arbitrary Function in Joseph Fourier 1768–1830 Ed. and Annotated by I. Cambridge, MA: Grattan-Guinness MIT Press. p. 1807

[10] Khaidukov V, Landa E, Moser TJ. Diffraction imaging by focusing-defocusing: An outlook on seismic superresolution. Geophysics. 2004; **69**(6):1478-1490

[11] Decker L, Klokov A, Fomel S. Comparison of seismic diffraction imaging techniques: Plane wave destruction versus apex destruction. In: SEG Technical Program Expanded Abstracts 2013. Houston, Texas: Society of Exploration Geophysicists; 2013. pp. 4054-4059

[12] Klokov A, Fomel S. Seismic diffraction imaging, one migration dip at a time. In: SEG Technical Program Expanded Abstracts 2013. Houston, Texas: Society of Exploration Geophysicists; 2013. pp. 3697-3702

[13] Bashir Y, Ghosh D, Sum C. Preservation of seismic diffraction to enhance the resolution of seismic data. In: SEG Technical Program Expanded Abstracts 2017. Houston, Texas: Society of Exploration Geophysicists; 2017. pp. 1038-1043

[14] Kozlov E, Barasky N, Korolev E, Antonenko A, Koshchuk E. Imaging scattering objects masked by specular reflections. In: SEG Technical Program Expanded Abstracts 2004. Denver, Colorado: Society of Exploration Geophysicists; 2004. pp. 1131-1134

[15] Landa E, Keydar S. Seismic monitoring of diffraction images for detection of local heterogeneities. Geophysics. 1998;**63**(3):1093-1100

[16] Spitz S. Seismic trace interpolation in the FX domain. Geophysics. 1991; **56**(6):785-794

[17] Nizarul O, Hermana M, Bashir Y, Ghosh DP. Improving thin bed identification in Sarawak Basin field

using short time Fourier transform half Cepstrum (STFTHC) method. In: IOP Conference Series: Earth and Environmental Science. IOP Publishing. 2016;**30**(1):012003

[18] Bashir Y, Ghosh DP, Alashloo SYM, Sum CW. Effect of frequency and migration aperture on seismic diffraction imaging. IOP Conference Series: Earth and Environmental Science. 2016;**30**(1):12001

[19] Berkovitch A, Belfer I, Hassin Y, Landa E. Diffraction imaging by multifocusing. Geophysics. 2009;**74**(6): WCA75-WCA81

[20] Moser TJ, Howard CB. Diffraction imaging in depth. Geophysical Prospecting. 2008;**56**(5):627-641

[21] Gray SH, Etgen J, Dellinger J, Whitmore D. Seismic migration problems and solutions. Geophysics. 2001;**66**(5):1622-1640

[22] Bleistein N. Two-and-one-half dimensional in-plane wave propagation*. Geophysical Prospecting. 1986;**34**(5):686-703

[23] Pusey LC, Vidale JE. Accurate finite-difference calculation of WKBJ traveltimes and amplitudes. In: SEG Technical Program Expanded Abstracts 1991. Houston, Texas: Society of Exploration Geophysicists; 1991. pp. 1513-1516

[24] Fomel S. Applications of plane-wave destruction filters. Geophysics. 2002; **67**(6):1946-1960

[25] Song X, Fomel S, Ying L. Lowrank finite-differences and lowrank Fourier finite-differences for seismic wave extrapolation in the acoustic approximation. Geophysical Journal International. 2013;**193**(2):960-969

[26] Fomel S, Ying L, Song X. Seismic wave extrapolation using lowrank symbol approximation. Geophysical Prospecting. 2013;**61**(3):526-536

[27] Fomel S. Three-Dimensional Seismic Data Regularization (Doctoral dissertation, Stanford University). Citeseer; 2000

[28] Claerbout JF. Earth Soundings Analysis: Processing Versus Inversion. Vol. 6. Massachusetts, USA: Blackwell Scientific Publications Cambridge; 1992

[29] Decker L, Janson X, Fomel S. Carbonate reservoir characterization using seismic diffraction imaging. Interpretation. 2014;**3**(1):SF21-SF30

[30] Versteeg R. The Marmousi experience; velocity model determination on a synthetic complex data set. The Leading Edge. Sep. 1994; **13**(9):927-936

[31] Bashir Y, Ghosh DP, Sum CW, Mahgoub M. Depth imaging using innovative algorithm for high-resolution seismic. In: RDPETRO 2018: Research and Development Petroleum Conference and Exhibition, Abu Dhabi, UAE, 9–10 May 2018. American Association of Petroleum Geologists, Society of Exploration Geophysicists, European Association of Geoscientists and Engineers, and Society of Petroleum Engineers. June 2018. pp. 20-23

[32] Bashir Y, Ghosh DP, Janjuhah HT, Sum CW. Diffraction enhancement through pre-image processing: Applications to field data, Sarawak basin, East Malaysia. Geosciences. 2018; **8**(2):74

[33] Bashir Y, Ghosh DP, Sum CW. Influence of seismic diffraction for high-resolution imaging: Applications in offshore Malaysia. Acta Geophysica. 2018;**66**(3):305-316

[34] Bashir Y, Ghosh DP, Weng Sum C, Janjuhah HT. Diffraction enhancement through pre-image processing:

Applications to field data, Sarawak Basin, East Malaysia. Geosciences. 2018; **8**(2):74

[35] Bashir Y, Ghosh DP, Sum CW. Diffraction amplitude for fractures imaging & hydrocarbon prediction. Journal of Applied Geology and Geophysics. 2017;**5**(3):50-59

[36] Claerbout J, Brown M. Two-dimensional textures and prediction-error filters. In: 61st EAGE Conference and Exhibition, Helsinki, Finland. 7–11 June 1999

[37] Claerbout J. Multidimensional recursive filters via a helix. Geophysics. 1998;**63**(5):1532-1541

[38] Fomel S, Claerbout JF. Multidimensional recursive filter preconditioning in geophysical estimation problems. Geophysics. 2003; **68**(2):577-588

Chapter 2

Radiometric Mapping of Hydrothermal Alterations in Isla Isabel, Mexico

Román Alvarez and Gerardo Figueroa

Abstract

Isla Isabel is a small island of volcanic origin about 29 km west of Nayarit's coast, Mexico, exhibiting phreatomagmatic activity (maars). In a nearby (8 km), ocean located, exploration well reaching 3157 m geothermal gradients of 73°C/km were recorded, which are among the largest in the world. These features, plus gravimetric anomalies in the area, suggest that the region may become an important source of geothermal energy. Direct hydrothermal alteration mapping in the island is difficult since it is a National Park and a bird sanctuary where various bird species thrive. We used remote sensing techniques to identify hydrothermally altered sections in the island, including radiometric spectrum determinations in the 350–1050 nm wavelength interval and Landsat 8 satellite imagery. The island is mainly composed of basalt but large portions contain vegetation, posing a challenge to satellite hydrothermal alteration detection. Four pixel-classification schemes are used for oxides, hydroxyls, and vegetation; shallow underwater basalt flow detection was possible.

Keywords: Isla Isabel, Mexico, hydrothermal alteration, phreatomagmatic activity, oxide and hydroxyl detection, radiometric spectra

1. Introduction

Isla Isabel is an island of volcanic origin located ~29 km west of the coast of Nayarit State and ~70 km east of Islas Tres Marias archipelago in the Mexican Pacific; its geographic location is 21°50.847'N, 105°53.000'W (**Figures 1** and **2**). The geologic composition of the island is basaltic with various explosion craters (maars) of Surtseyan (phreatomagmatic) origin scattered in the surface [1]. The island dimensions are 1.8 km N-S and 1.32 km E-W.

A regional study around Isla Isabel showed that the island represents the emerged portion of a larger body preliminarily identified as a laccolith [2]. Over the average depth of 100 m of the continental platform in which the island is located, a topographic bulge rises to the surface; its dimensions are 20 km in the NW-SE direction and 17 km in the perpendicular direction. This bulge has been assumed to be built by successive magmatic intrusions that are deflected sideways by the sediments; the island represents the only emerged portion of this laccolith. In addition to the regional study, local geophysical surveys have been made on the emerged portion. Interpretation of magnetic and gravity results on the island's surface suggests the existence of diatremes associated with some of the explosion craters, as well as identification of a growth boundary about 800 m deep, from which apparently originated

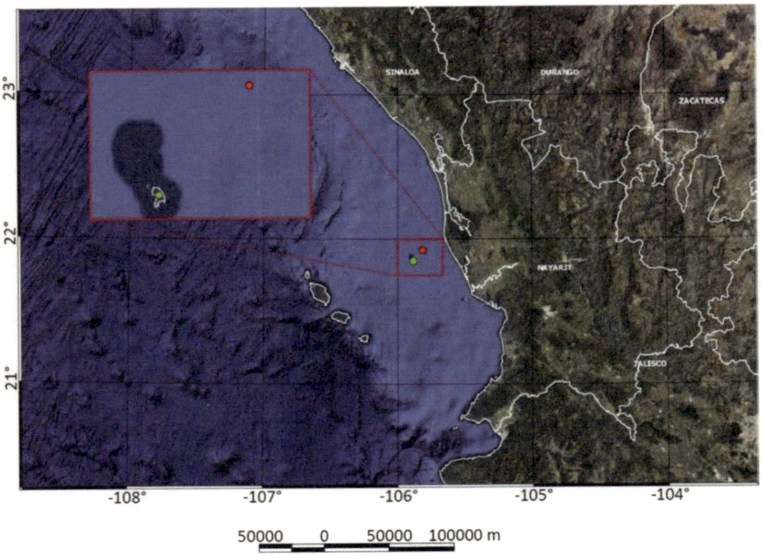

Figure 1.
Location of Isla Isabel (green dot) and exploration well Huichol I (red dot). Tres Marías islands are located SW of Isla Isabel.

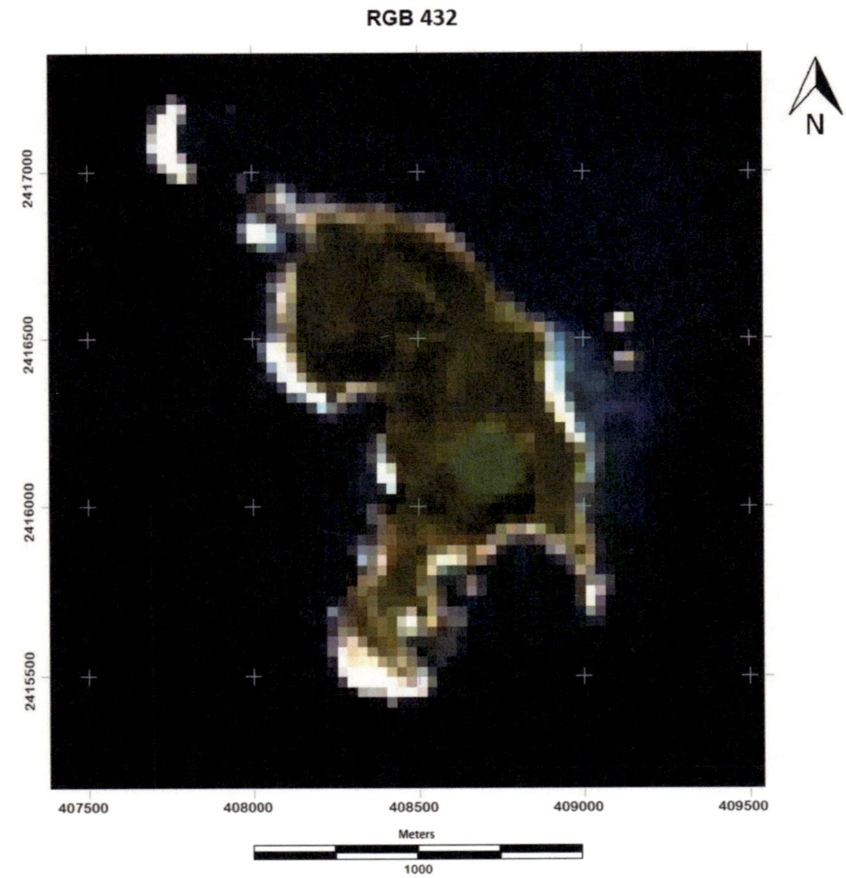

Figure 2.
Landsat 8 image Path/Row 31/45, code LC80310452014002LGN00 obtained on February 13, 2014. The image (RGB composite 4-3-2) is shown after image crop out, and atmospheric and radiometric corrections were made. Spatial resolution is 30 m. The fuzzy appearance is due to the scale used, which begins to show individual pixels.

one of the latest constructional, basaltic episodes [3]. Alkali basalts have transported mafic and ultramafic peridotite xenoliths to the surface; sequences of dark-red, near-vent scoria deposits, and lava flows in the tidal zone have also been reported [1].

Huichol I exploration well was drilled about 8 km NE of the island (**Figure 1**) penetrating 3157 m, with diabase intersected in the deepest 250 m. Geothermal gradients reached the values of 73°C/km, some of the largest measured in the world [4], highlighting the anomalously high heat flux in the area.

The above observations suggest that the island and its surroundings deserve geological and geophysical studies to evaluate its geothermal potential. In the present study, we focus on hydrothermal alteration detection and mapping in the island, with remote sensing techniques.

2. Radiometric spectra

The radiometric aspects of the island materials are approached from direct spectrum determinations with a radiometer and from the numerical analysis of the selected Landsat 8 image. To obtain the reflectance of the sampled surfaces, we used an Ocean Optics USB2000+VIS+NIR radiometer with a bandwidth ranging from 350 to 1050 nm. These reflectances are then compared to those of the corresponding bands of the Landsat 8 image. The rocks measured radiometrically were analyzed by X-ray fluorescence obtaining their elemental compositions, which also allowed for comparison with the spectral signature of the corresponding database of the US Geological Survey [5].

Landsat 8 images have 11 bands; of them, we used coastal/aerosol, blue, green, red, Near Infra-Red (NIR), and the 2 Short Wavelength Infra-Red (SWIR) bands. The atmospheric correction was performed with the Histogram Minimum Method [6], and resampling of the digital numbers was made to fit into the 8-bit radiometric resolution format. The spatial resolution was not changed. **Figure 2** shows a color composition (RGB-432) of the image after preprocessing was accomplished. At this scale, the individual pixels of 30 × 30 m begin to show, giving the image a fuzzy appearance.

2.1 Radiometric stations

Ten stations were sampled across the island; at each station, several radiometric measurements were made, and rock samples were collected for subsequent composition analyses. **Figure 3** shows the location of each station and **Table 1** summarizes pertinent information. Owing to space reasons, data obtained in some stations will be omitted in this report.

2.1.1 Station 1

This station corresponds to a basalt outcrop, and the reflectance spectra are shown in **Figure 4**. The four curves follow the same tendencies, with the same variations in percentage reflectance, indicating a consistent response at the station. The different reflectance baselines are attributed to different collection angles. Reflectance is fairly low in Station 1; the four spectra show a wide maximum between 440 and 640 nm.

2.1.2 Station 2

This station contains two different lithological units. The first is a basalt outcrop, and the second corresponds to white pumice. **Figure 5** shows the outcrops and the

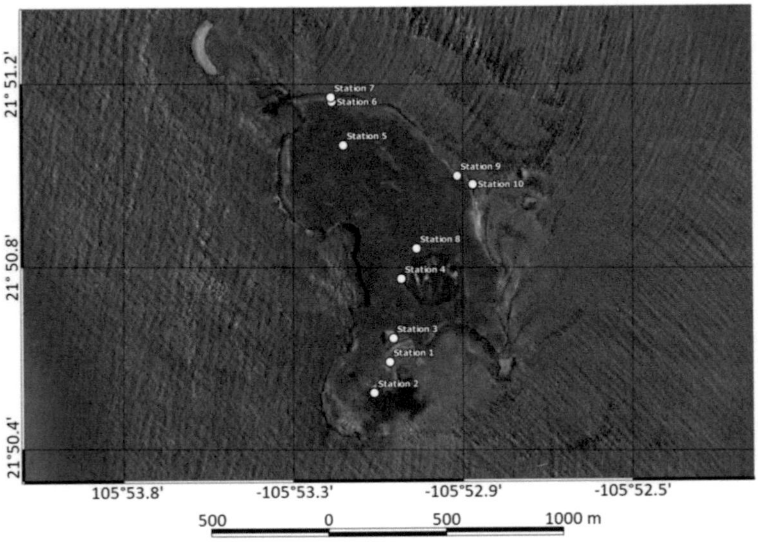

Figure 3.
Image of Isla Isabel composed of Google Earth images from June 23, 2015. The location of the radiometric stations corresponds to the white dots.

Station	Easting	Northing	Samples	Type
1	408,543	2,415,752	4	Basalt
2-1	408,476	2,415,621	4	Basalt
2-2			2	White tuff
3-1	408,558	2,415,852	4	Soil
3-2			4	Vegetation
4-1	408,594	2,416,106	4	Basalt
4-2			2	Basalt
5	408,349	2,416,672	1	Basalt
6	408,302	2,416,856	1	Basalt
7	408,298	2,416,874	4	Basalt
8	408,659	2,416,232	3	Vegetation
9	408,659	2,416,540	3	Basalt
10-1	408,834	2,416,505	3	Tuff
10-2	408,659	2,416,232	3	Basalt

Table 1.
Station location in UTM coordinates, number of spectral samples, and corresponding rock type.

measuring equipment, and **Figures 6** and **7** show the radiometric responses of basalt and the white tuff. As in Station 1, the basalt exhibits consistent radiometric responses in the four determinations. However, the response differs slightly from that of basalts in Station 1, since they show a plateau between 500 and 600 nm, where the former showed a broad reflectance peak.

The white pumice, as expected, shows radically different spectra from that of basalt. These determinations also show consistency between them, being characterized by a monotonous increase in reflectivity between 360 and 700 nm which evoque the response of a dry, bare soil followed by rapid variations at the end of the spectrum.

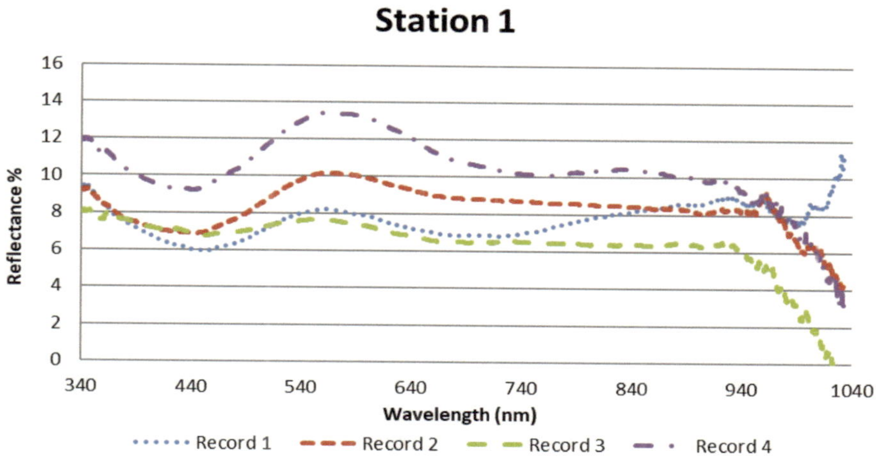

Figure 4.
Four radiometric records obtained at Station 1 in Isla Isabel. The station location is shown in **Figure 3**, *and its coordinates appear in* **Table 1**. *The rock composition corresponds to basalt.*

Figure 5.
Measuring basalt with the spectrometer; the PC control and record measurements. At Station 2, two types of materials were sampled: Basalt and white tuff.

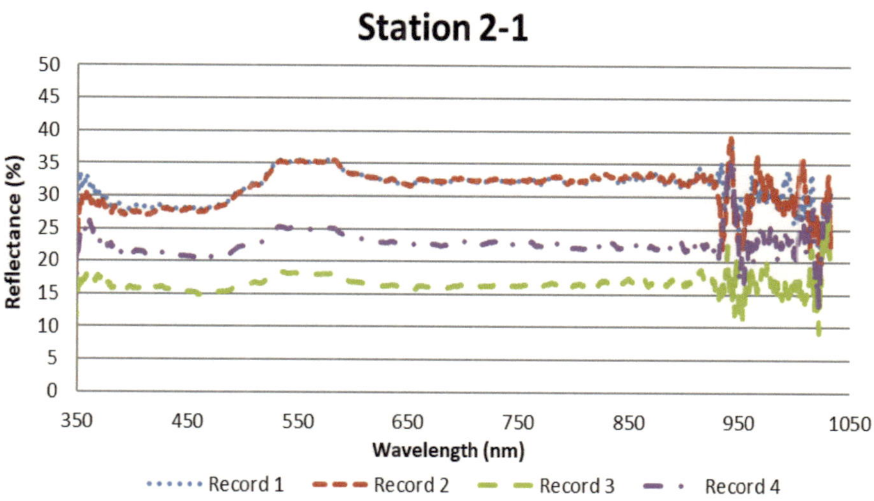

Figure 6.
Four records of the radiometric response of basalt at Station 2-1.

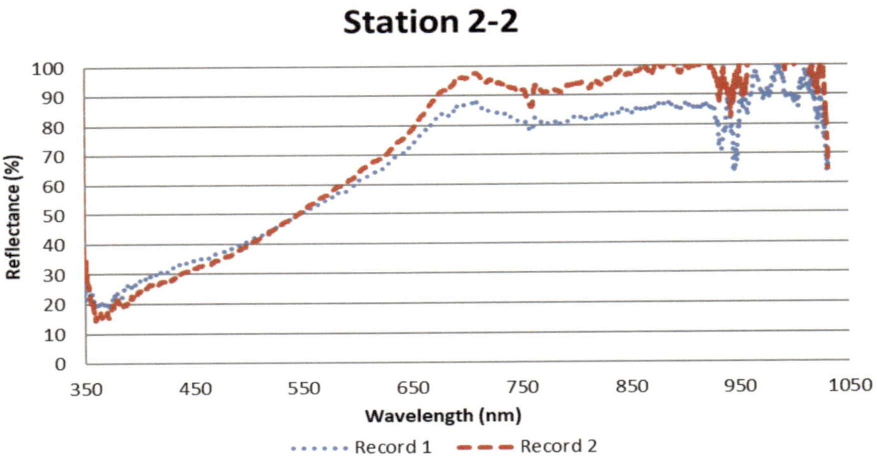

Figure 7.
Radiometric response of white pumice at Station 2-2.

2.1.3 Station 4

This station is located on the margin of Lago Crater (also known as Laguna Fragatas) as shown in **Figure 3**; according to [1], this structure is the only, fully preserved, explosion crater in the island. The exposed rocks on the lake's margin are basalts altered either by the fluctuating water level (**Figure 8**) or by the chemical action of birds' excrement. The latter mechanism appears more viable since birds are abundant in the island; Islote Pelon, slightly north of the island, is an extreme case of these alterations, being completely covered by bird feces. The white surfaces of the rocks induced saturation in the radiometer making it difficult to obtain valid readings. The spectra from Station 4-1, shown in **Figure 11**, were obtained pointing to dark spots of the chemically altered rock samples. We observed a systematic tendency for reflectance to diminish as the wavelength increases; the large reflectance contribution to the lower part of the spectrum probably arises in dispersed radiation at the white surfaces of the rocks as well as at the lake's surface.

Although distorted by the dispersion effect, two of the signatures clearly show a similar increment in reflectance between 500 and 600 nm as those shown by basalt

Figure 8.
Chemically altered basalts on the margin of Lago Crater. We attribute this alteration to the chemical effect of bird feces.

in Station 1 and Station 2-1. Between 690 and 900 nm, it also shows the vegetation signature, compare to that of Station 8 (**Figure 12**).

2.1.4 Station 5

This station is in a bushy area where vegetation height ranges between 2 and 3 m. Many birds nest in the area, and the outcropping rocks have been exposed to their excrements during long periods. The surface of the sample chosen for the measurement was under such conditions. In **Figure 9**, a hand specimen from the same outcrop is shown; its exterior is chemically altered, while its interior shows the dark color of unaltered basalt. The radiometric response is shown in **Figure 10**. Reflectance is flat from 450 to 700 nm and quite similar to the response of the altered surface at Station 4-1, Record 1 (**Figure 11**) in the same interval.

Figure 9.
Basalt hand specimen collected at Station 5. The white surface has been exposed and chemically altered, while the fresh portion appears black.

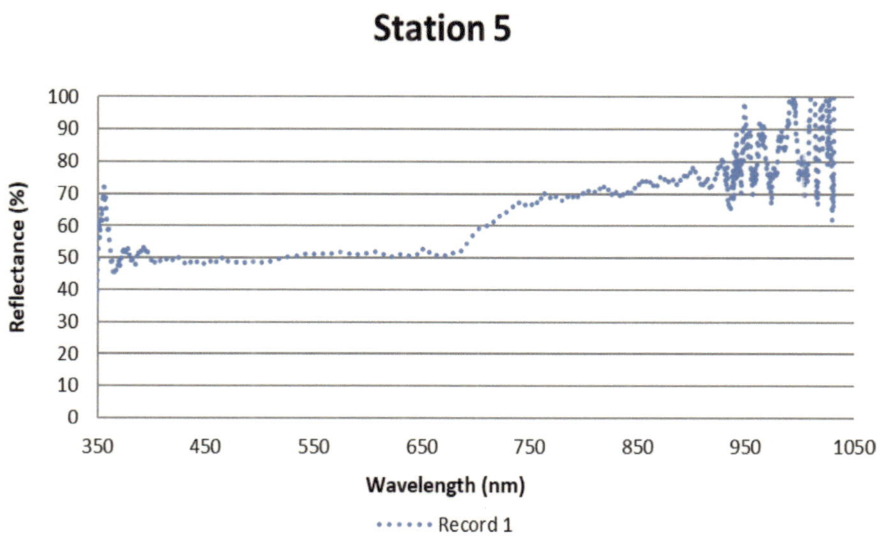

Figure 10.
Radiometric response at Station 5. A combination of shade and chemically altered surface may correspond to the flat response between 400 and 700 nm. The reflectance increment between 700 and 950 nm is attributed to the effect of vegetation surrounding the station.

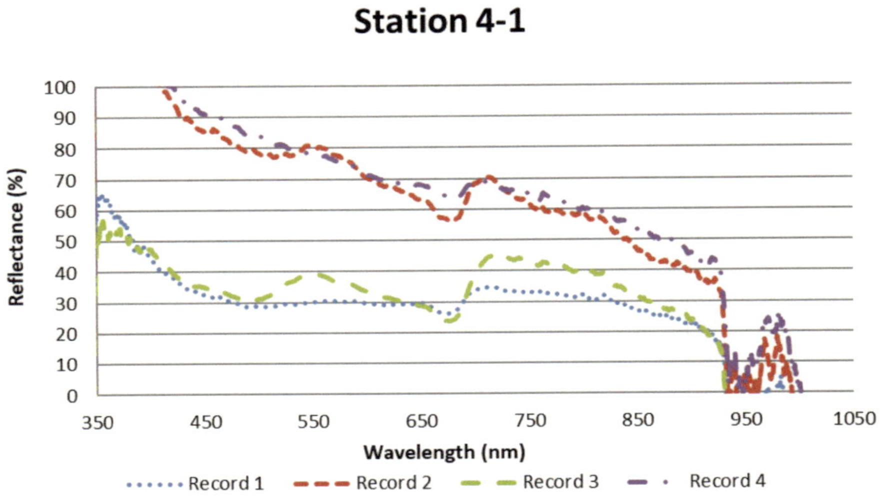

Figure 11.
Spectral signature of chemically altered basalt in the shore of Lago Crater. It was obtained pointing to the darker portions of the rock. Reflectance increase in the lower part of the spectrum is attributed to dispersion effects.

Although no direct testing for the effects of bird's feces on basalt is available, bird droppings are known to chemically alter hard surfaces such as concrete [7]. These authors analyzed samples of concrete using scanning electron microscopy for morphological properties and X-ray spectrophotometry for surface elements, finding that the components of bird excrement penetrate microcracks causing it to become smoother. They also found that concrete materials exposed to bird feces lost weight more than the control samples.

2.1.5 Station 8

This station is located on the rim of Lago Crater; the interest here was to obtain a clean vegetation record from the top of the rim. Three records are shown in **Figure 12**; they neatly show the radiometric signature of vegetation, with a reflectance peak centered at 550 nm and a large reflectance in the 700–950 nm NIR region, typical of healthy vegetation [8]. The records differ in percent reflectance but basically maintain the same

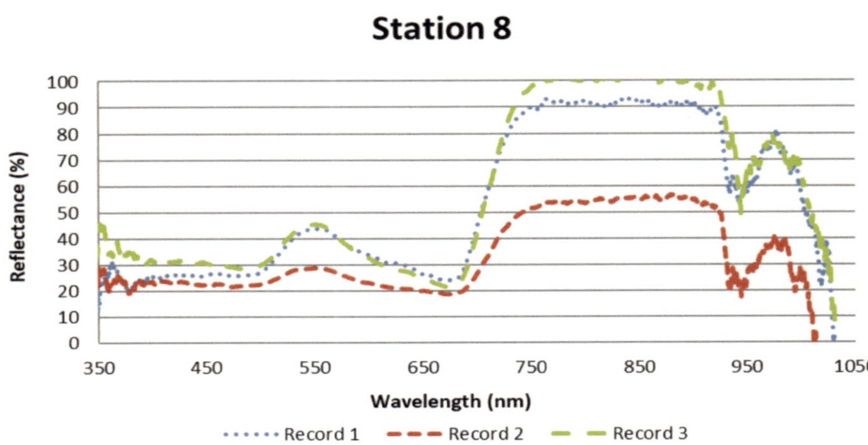

Figure 12.
Station 8 is located on the top of the north rim of Lago Crater. The objective was to obtain the radiometric signature of vegetation around the crater.

spectral signature. The reflectance peaks centered at 550–980 nm are well defined and differ from the response of the geologic materials obtained in other stations.

2.1.6 Station 9

Located on the eastern portion of the island, and this station corresponds to basalt without chemical alteration. Notice the low reflectance values (**Figure 13**), indicating the strong absorption of this rock. The larger reflectance noted in previous basalt samples in the 500–600 nm range is also present here, although somewhat subdued by the strong absorption. The rapid variations in the NIR range show large amplitudes in the three records.

2.1.7 Station 10-2

This station corresponds to basalt not affected by chemical alteration. The records are shown in **Figure 14**, they show coherency among themselves, although Record 1 shows an anomalously high baseline reflectance, which we attribute to calibration effects. Record 1 shows rather high levels of reflectance with a well-defined plateau between 500 and 600 nm, as previously noted for other basalt samples. This feature is also present, although somewhat subdued, in Records 2 and 3. Between 350 and 500 nm, the latter shows a monotonous decrease in reflectivity that is not present in Record 1; they are attributed to the dispersion effect previously observed in other samples in this study. The NIR responses of the three records are quite similar and show various localized radiation absorptions.

2.1.8 Summary

Summarizing these radiometric results, we conclude that basalts present a distinct signature throughout the island, showing low reflectance, usually below 20%, that is distorted in some stations by alterations and/or the influence of disperse radiation. The calibration procedure performed prior to each measurement, and

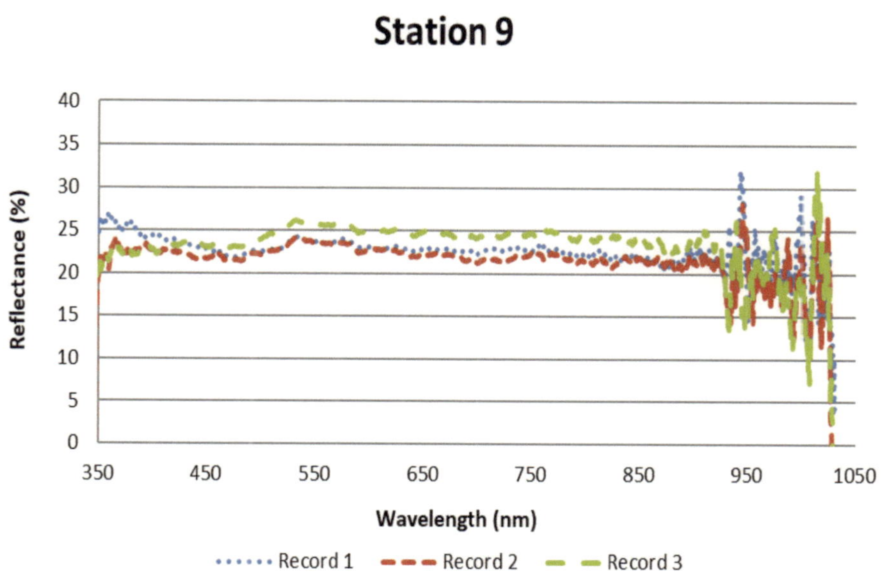

Figure 13.
Station 9 is located on the eastern portion of the island where unaltered basalts are clearly exposed. This station and Station 1 showed the lowest reflectance values in the island.

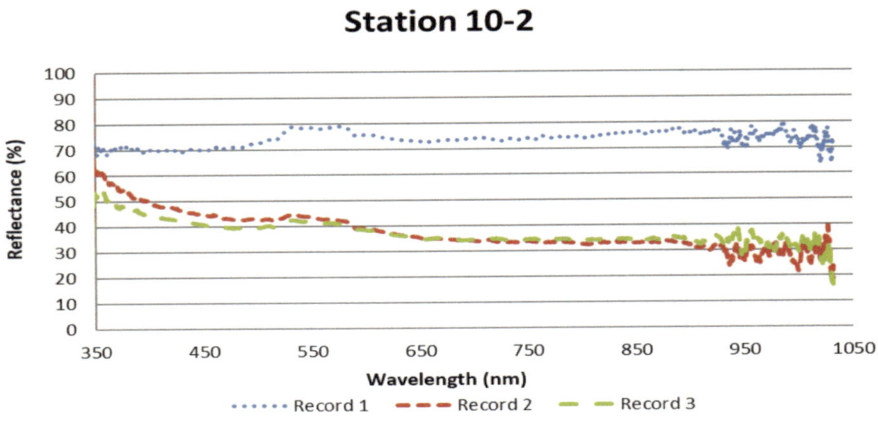

Figure 14.
Radiometric signatures of basalt at Station 10-2. These basalts did not present signs of chemical alteration. The baselines show larger values than those previously observed in basalt.

the angle of measurement are most likely responsible for the baseline variations between measurements. Without considering the NIR band, we can establish a plateau reflectance response in the 500–600 nm bandwidth followed by a flat region between 650 and 900 nm, as the characteristics of the basalt response in this group of measurements (**Figures 4, 6,** 7, and **14**).

3. Comparison of satellite and radiometric data

To better evaluate the type of radiometric departures between in situ measurements, which sample small, specific locations at an observation station, and satellite determinations which necessarily involve averaging the radiometric responses of the various materials in each pixel, we performed a comparison between them. We used the least noisy spectrum for each station as well as the values of the Landsat 8 pixel that contains the corresponding station, obtaining the values of the digital numbers of each band: B1, B2, B3, B4, and B5; these values were converted to a 0–100% reflectance scale and plotted with the corresponding spectral signature (**Figures 15–17**). The SWIR bands do not overlap the radiometer's range.

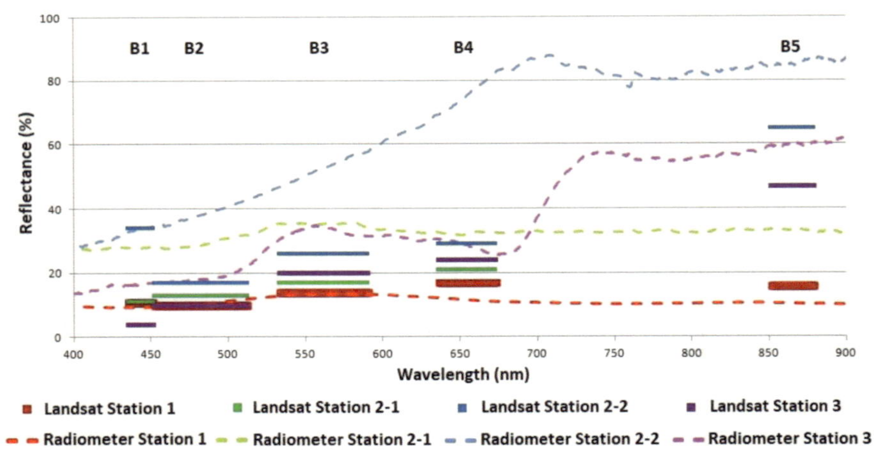

Figure 15.
Reflectance comparison between Landsat 8 pixels and radiometric stations. Stations 1, 2-1, 2-2, and 3. The band coverage is shown as colored, continuous lines, while dashed lines correspond to the radiometric determinations.

In **Figure 15**, the low reflectance obtained with the radiometer in Station 1 corresponds well with the values obtained from the pixel of the Landsat 8 image. In Station 2-2, the radiometer readings overshoot those of Landsat 8 but maintain the general trend. Stations 2-1 and 3 correspond well with the satellite radiometric responses. In **Figure 16**, Station 4 departs considerably from the satellite band behavior. Although Stations 5, 6, and 7 conform to the general trends of the satellite measurements, reflectance is considerably larger.

In **Figure 17**, radiometric Stations 8, 9, and 10-2 show a behavior similar to the Landsat 8 bands, although displaced and amplified. Station 10-1 departs from the satellite bands owing mostly to the dispersion effect clearly shown below 700 nm.

We conclude that there is a fair agreement between the satellite and the in situ measurements. With the ground truth of the radiometric stations, we can now proceed to analyze the satellite data.

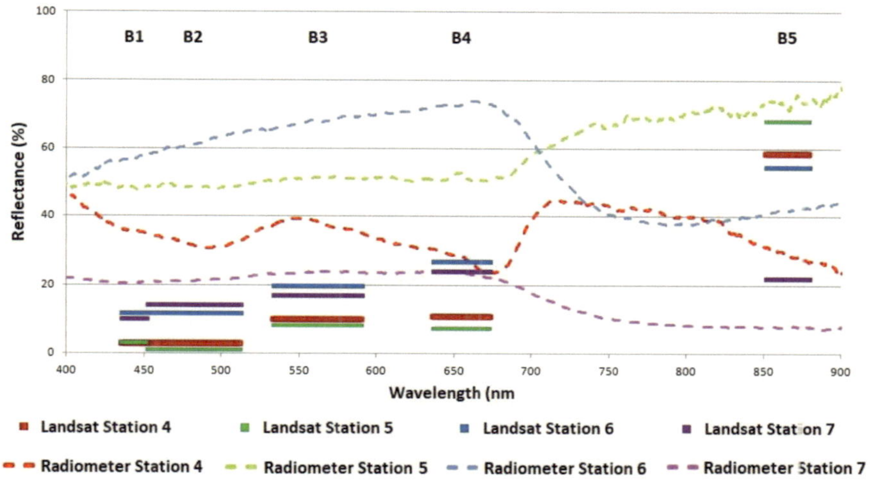

Figure 16.
Reflectance comparison between Landsat 8 pixels and radiometric stations. Stations 4, 5, 6, and 7.

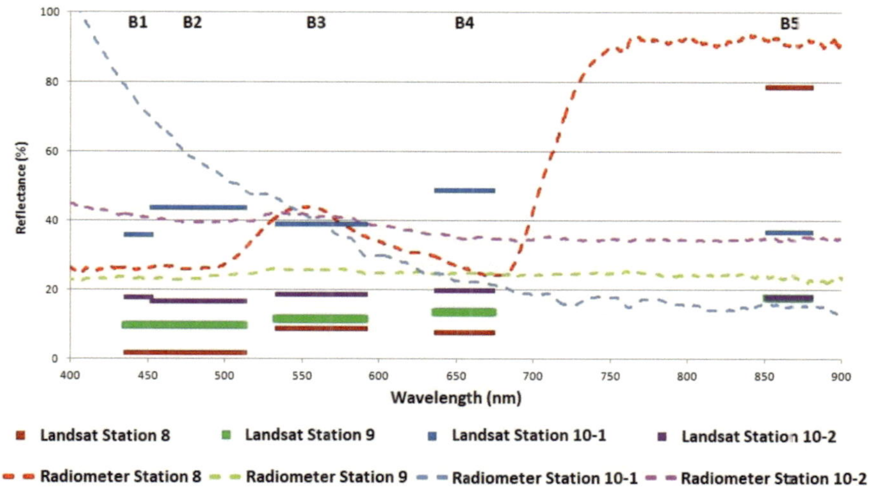

Figure 17.
Reflectance comparison between Landsat 8 pixels and radiometric stations. Stations 8, 9, 10-1, and 10-2.

4. Geochemistry

To properly associate radiometric response to rock composition, we need to perform a geochemical analysis on the samples where reflectance was measured, particularly on those identified as basalts, since they are the most common. Ten geochemical analyses throughout the island have been published [1]; we deemed sufficient to analyze only three additional ones, on samples collected at Stations 1, 2, and 5.

The elemental compositions obtained by means of X-ray fluorescence (XRF) appear in **Table 2**. Their basic composition is readily appreciated from their low silica content. After data normalization, the results were processed with the IGPET version 2005 software [9] to get the total alkali silica (TAS; [10]) and Alkali FeOMgO (AFM; [11]) diagrams are shown in **Figures 18** and **19**. Results are quite similar to the elemental data for the ten whole-rock samples across the island reported in [1].

Component	ISA-01	ISA-02	ISA-05
SiO_2	46.421	43.146	47.316
TiO_2	1.456	1.566	1.508
Al_2O_3	14.64	15.34	15.35
Fe_2O_3t	9.357	9.315	9.052
MnO	0.159	0.15	0.15
MgO	10.107	9.226	10.06
CaO	11.25	10.108	9.869
Na_2O	2.749	3.283	3.172
K_2O	1.341	1.501	1.526
P_2O_5	1.056	1.671	0.856
PXC	1.36	1.59	1.04
Total	99.896	96.896	99.899

Table 2.
Elemental composition of three samples from Isla Isabel from X-ray fluorescence (wt%); they correspond to basalt.

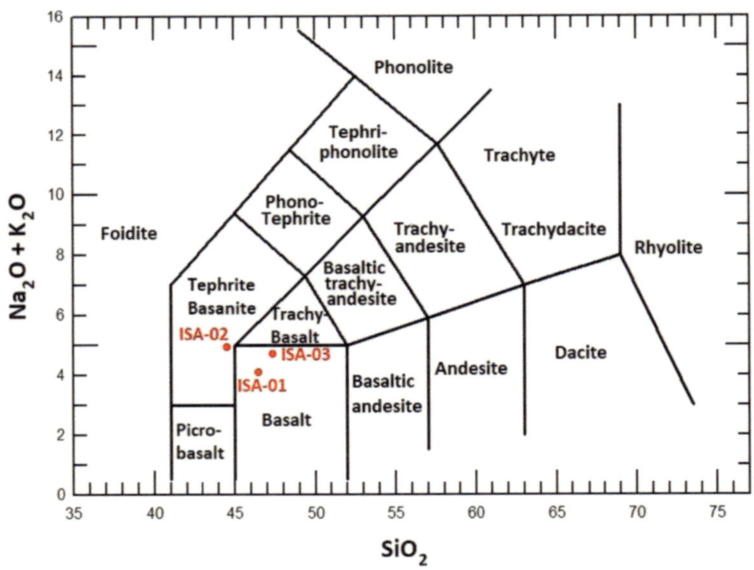

Figure 18.
Total alkali silica (TAS) diagram of samples ISA-01-03 (red) classified as basalt and tephrite-basanite from the elemental compositions in **Table 2.**

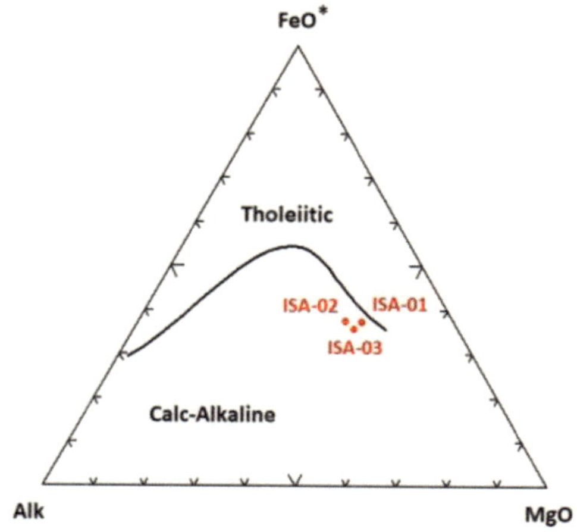

Figure 19.
Alkali FeOMgO (AFM) diagram of samples ISA-01-03 (red) from the elemental compositions in **Table 2**.

Mineral content of basalts: groundmass generally of pyroxene (augite), plagioclase and olivine, possibly with minor glass; if porphyritic, the phenocrysts will be any of olivine, pyroxene or plagioclase. Silica (SiO_2) content—45–52%.

5. Mapping hydrothermal alterations

The date of **Figure 2** corresponds to the dry season, where vegetation influence is minimized; the scene is devoid of clouds. Landsat 8 images contain 11 bands; in this processing, we used coastal/aerosol, blue, green, red, NIR, and 2 of the SWIR bands [12].

Atmospheric correction was performed with the Histogram Minimum Method (e.g., [13]). The radiometric correction was performed resampling the image's digital numbers to fit the 8-bit radiometric resolution; the spatial resolution was not modified. The image with these corrections was presented above in **Figure 2**.

To map regions of hydrothermal alteration, we focus on enhancing the spectral response of their typical mineralogical contents. Three major groups characterize hydrothermal alterations (e.g., [14]), hydroxyls (clays and micas), iron minerals (hematite, goethite, and jarosite), and hydrated sulfates (chalk and alunite). Their spectral signatures appear in **Figure 20**.

The identification of hydrothermal alterations is approached with four methods, to be discussed below, applied to the satellite image: Band Ratio, Principal Component Analysis (PCA), the Crosta Technique (Crosta), and the Fraser Technique (Fraser). Results are compared and discussed to optimize the hydrothermal identification.

5.1 Band ratio

Enhancement of hydrothermal alterations is performed making the ratio of the satellite bands that better characterize them. Bands B4/B2 for the oxides, and bands B6/B7 for the hydroxyls. The ratio B5/B4 is included to represent vegetation in the color composition representation of these components, as shown in **Figure 21**.

The digital numbers resulting from the band ratios are rescaled to cover the digital values from 0 to 255 for each ratio. A better visualization is obtained when the false color image is classified; we obtain the percentage of pixels belonging to each class: oxides, hydroxyls, and vegetation, as well as their combinations, and a marine water class, which define eight classes. K-means clustering [16] was used to partition the n-observations into k-clusters, in which each observation belongs to the cluster with the nearest mean, resulting in a prototype of the cluster. The result of applying this algorithm to the image in **Figure 21** is shown in **Figure 22**, where the classes have been color coded.

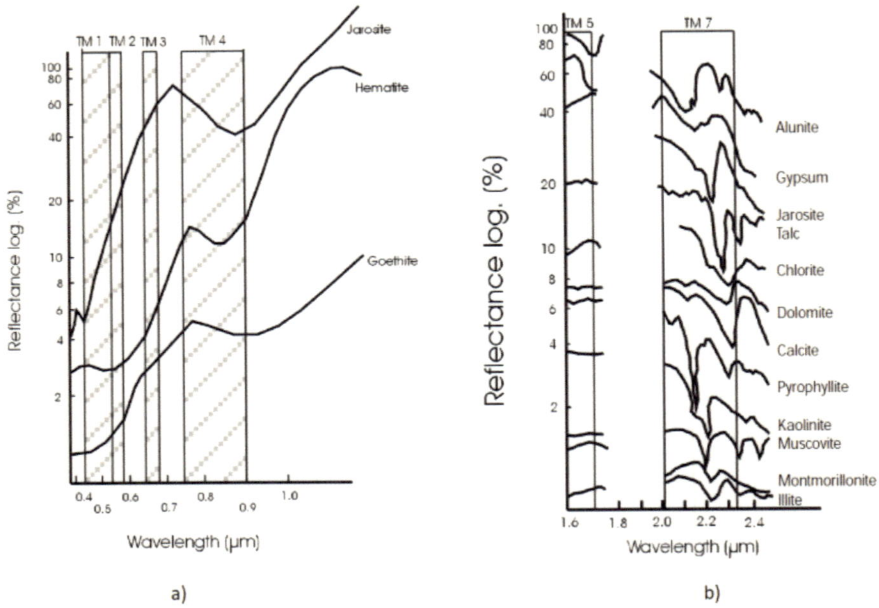

Figure 20.
Spectral signatures of hydrothermal alteration minerals. (a) Oxides and (b) hydroxyls. Adapted from [15].

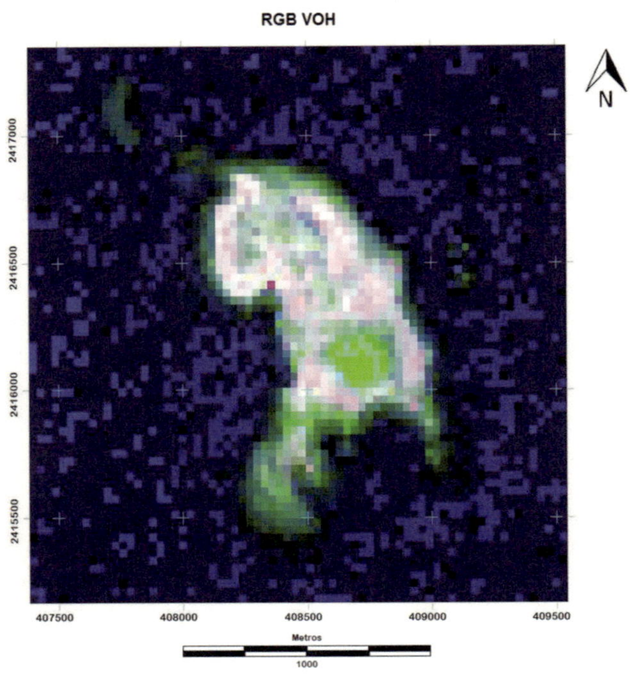

Figure 21.
RGB composition of the band ratio process: vegetation (B5/B4, red), oxides (B4/B2, green), and hydroxyls (B6/B7, blue). Vegetation pixels and oxide pixels are abundant, while a few cyan colors are observed, which represent the combined response of oxides and hydroxyls to hydrothermal alterations. The fuzzy appearance is due to the scale used, which begins to show individual pixels.

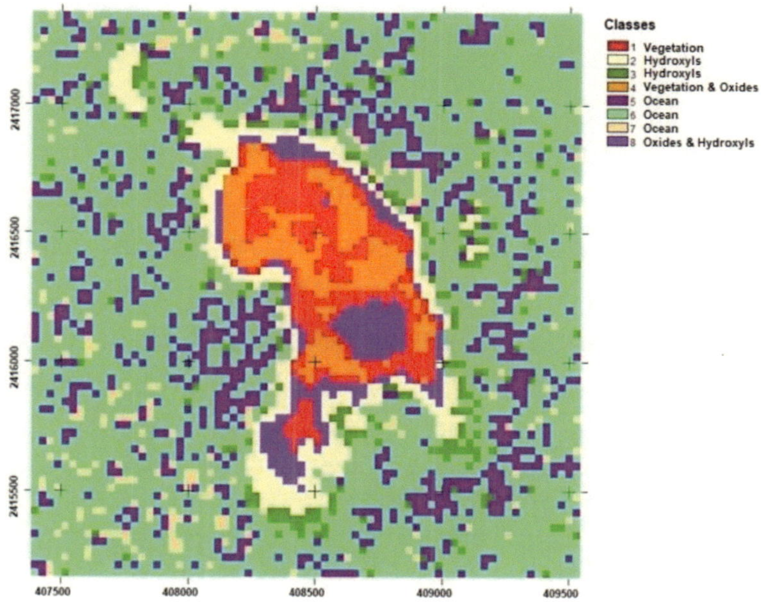

Figure 22.
Classification resulting of applying the K-means algorithm to the band ratio and RGB image in **Figure 21**.

Type	Number of pixels	Class
Ocean	3992	5–7
Vegetation	—	—
Oxides	—	—
Hydroxyls	562	2, 3
Vegetation & oxides	508	4
Vegetation & hydroxyls	271	1
Oxides & hydroxyls	194	8

Table 3.
Type, class, and the number of pixels in each class in the classified image of **Figure 22**, *obtained with the command "**Area**" in IDRISI.*

As can be appreciated from these results, oxides and hydroxyls (Class 8) are confused with the water of Lago Crater. From **Table 3**, we can see no pixels associated with vegetation or oxides, indicating the need for a better class identification scheme.

5.2 Principal component analysis

Principal component analysis (PCA) is a procedure to decorrelate a set of original variables by means of orthogonal transformations [17]. The principal components are linear combinations of the original variables, and it is expected that only the first ones contain the largest variability, obtaining a decrement in the data dimensions. IDRISI [18] contains a module to calculate the resulting matrix; each band of the cropped image (**Figure 2**) is loaded into the program. **Table 4** shows the result of the calculation. As an example of the use of the matrix, we recall the band ratio for the oxides: B4/B2. Across B2 and B4, we select the two most distant values from zero value (positive and negative), finding this condition in C4, B2, and B4; thus {B4/B2} = 0.100/−0.159. For hydroxyls, we used band ratio B6/B7, finding the above

	C1	C2	C3	C4	C5	C6	C7
B1	0.256965	0.887104	-0.348040	-0.138155	0.059819	-0.008275	0.056119
B2	0.306908	0.876144	-0.333635	-0.159081	0.011407	0.009315	0.036665
B3	0.647203	0.710474	-0.245058	0.054281	-0.115471	0.000692	0.004537
B4	0.662972	0.731901	-0.096873	0.100614	0.072249	-0.006020	-0.005010
B5	0.989940	-0.129653	-0.056608	0.000090	0.001800	0.000952	0.000014
B6	0.958162	0.191692	0.211512	-0.014765	-0.006699	-0.013415	-0.000511
B7	0.904145	0.324699	0.273529	-0.000734	0.0028420	0.047437	0.003909

Rows show the original bands and columns show the principal components. Blue, green, and red identify selections.

Table 4.
Weight matrix of the principal component analysis.

Type	Number of pixels	Class
Ocean	4241	5, 7
Vegetation	238	1
Oxides	197	2, 4
Hydroxyls	238	3
Vegetation & oxides	—	—
Vegetation & hydroxyls	215	8
Oxides & hydroxyls	127	6

Table 5.
Classification of pixel type obtained with the K-means algorithm for the PCA.

condition along C6. This criterion is not fulfilled for the ratio B5/B4 representing vegetation; to obtain a false color image for PCA, we introduce it separately.

A PCA representation of vegetation, oxides, and hydroxyls (RGB) similar to that in **Figure 22** is obtained (not shown). The K-means algorithm was also applied to that image, and the results appear in **Table 5**. Vegetation and oxide pixels are now present although the combined vegetation & oxides class could not locate any. The ocean classes show a new, large area NE of the island that is tentatively identified with the existence of basalt flows at the shallow ocean bottom. The classification result is shown in **Figure 23**.

5.3 Crosta technique

A variant of the PCA is the oriented principal components (OPC), also known as the Crosta Technique [19]. This technique consists of subjecting the specific bands of a given type, and bands not associated with it, to a PCA. A new PCA was calculated with the bands representative of vegetation (B4 and B5) and the band showing the least possible correlation, which in this case is Band 7. **Table 6** is similar to **Table 4**, but with only three components; the band ratio (B5/B4) can be

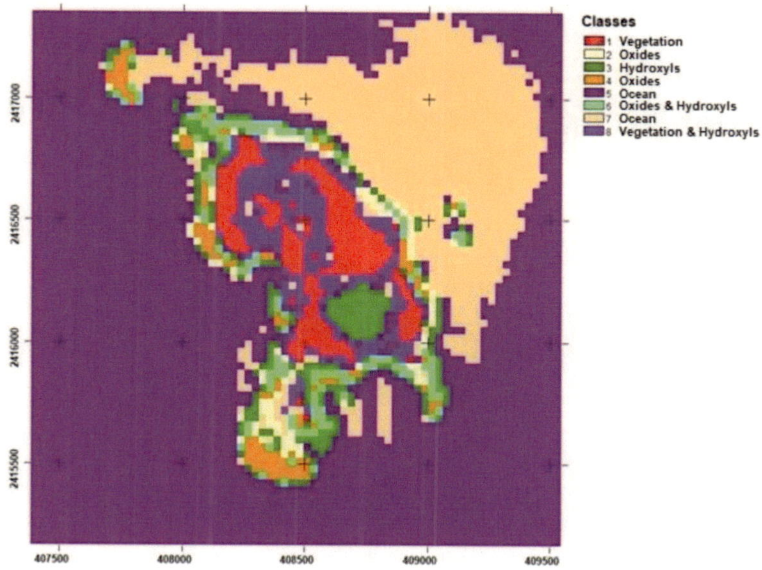

Figure 23.
The K-means algorithm was applied to the PCA. The class ocean (light brown) is observed near shcre and at a large lobe on the NE portion of the figure corresponding to the rim of the island where basalts are exposed at the shallow bottom.

	C1	C2	C3
B4	0.606464	0.863104	0.168687
B5	0.998631	-0.051582	0.008732
B7	0.863104	0.427635	-0.268663

Table 6.
Weight matrix for the Crosta technique applied to vegetation.

	C1	C2	C3	C4
B2	0.254797	0.902392	0.31147	0.154124
B4	0.610726	0.785016	0.040743	-0.095410
B5	0.998373	-0.055148	0.014464	0.000033
B7	0.864659	0.381888	-0.323275	0.044944

Table 7.
Weight matrix for the Crosta technique applied to oxides.

	C1	C2	C3	C4
B2	0.26655	0.660609	0.701797	-0.005234
B5	0.99402	-0.107101	0.021272	0.001074
B6	0.95101	0.297992	-0.080951	-0.015088
B7	0.891873	0.441765	-0.082842	0.050436

Table 8.
Weight matrix for the Crosta technique applied to hydroxyls.

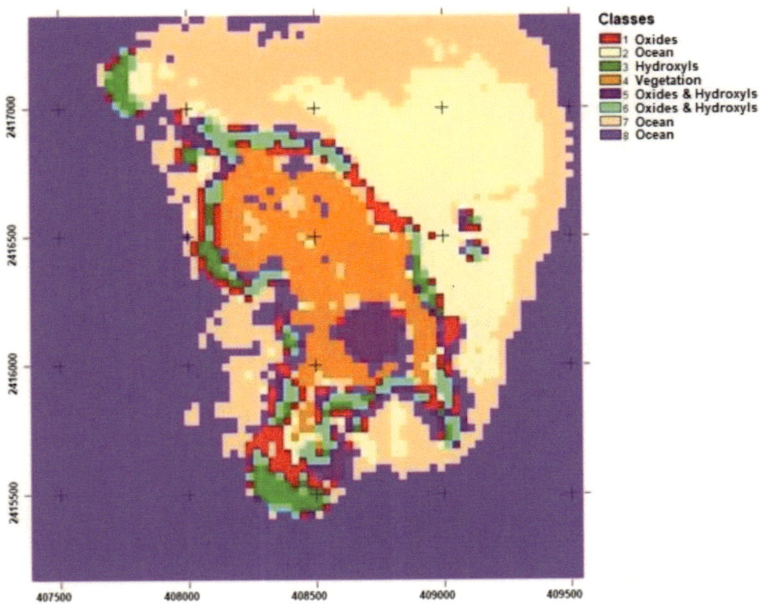

Figure 24.
The K-means algorithm was applied to the results of the Crosta calculation to obtain the number of pixels in each type. Two classes of ocean pixels are distinguished here, the one with the lighter tone corresponding better with the area of the clearly visible submerged basalts.

Type	Number of pixels	Class
Ocean	4332	2, 7, 8
Vegetation	396	4
Oxides	176	1
Hydroxyls	96	3
Vegetation & oxides	—	—
Vegetation & hydroxyls	—	—
Oxides & hydroxyls	256	5, 6

Table 9.
*Classification of pixel type obtained with the K-means algorithm for the PCA of the Crosta technique results (***Table 6***).*

obtained from column C2. Additionally, two independent analyses are performed for oxides and hydroxyls; for oxides, we take bands 2, 4, 5, and 7 to avoid mapping hydroxyls. The corresponding weight matrix appears in **Table 7**. For the hydroxyls, we used bands 2, 5, 6, and 7 to avoid mapping oxides. The corresponding weight matrix is shown in **Table 8**.

A RGB color composition can now be made with the principal components that represent hydroxyls, oxides, and vegetation (not shown), as previously done with the band ratios in **Figure 24**. The K-means algorithm was applied to it; the result is shown in **Table 9**.

Pixel identification resulting from the Crosta Technique for vegetation and the PCA results for oxides and hydroxyls show that hydroxyl identification improved with respect to the PCA results. We applied the K-means algorithm to this procedure to define the number of pixels that correspond to each type. Results appear in **Table 9** and **Figure 24**. This result shows class ocean (light brown) mainly around the island and in a large lobe NE of the island, interpreted as shallow-depth (≤10 m) regions that contain visible basalt flows at the bottom. With this algorithm, we obtained a larger number of pixels of oxides and hydroxyls than those obtained with the band ratio method. However, it still cannot properly identify regions of vegetation with oxides or hydroxyls.

Although this is an improved version of the band ratio method, we also observe that the Crosta algorithm has not been able to fully identify the different pixel types, as shown in **Table 7**. Nonetheless, this algorithm shows improvements with respect to the former since it has distinguished more pixels of the oxide-hydroxyl mixture (Classes 5, 6), which is the objective of this work. Additionally, it shows some yellow pixels representing the mixed type of oxides and vegetation, only present in the band ratio and in the Fraser technique, to be discussed later.

5.4 Fraser technique

This technique uses input bands as the band ratios that highlight the spectral characteristics of the materials of interest [20]. The idea is to separate the spectral differences of the materials, accomplished via the two resulting eigen vectors. In this process, a PCA analysis must be performed twice, one involves the band ratio

	C1	C2
Vegetation	0.982941	−0.183921
Oxides	0.890346	0.455285

Table 10.
Weight matrix for the Fraser technique applied to vegetation and oxides.

	C1	C2
Vegetation	0.986763	−0.162172
Oxides	0.708288	0.705924

Table 11.
Weight matrix for the Fraser technique applied to vegetation and hydroxyls.

for vegetation and the band ratio for the oxides. The second involves the band ratio for vegetation and the band ratio for hydroxyls. **Tables 10** and **11** show the corresponding results. To continue with the analysis, the associated RGB color composition image was prepared (**Figure 25**). Subsequently, the K-means algorithm is applied to the RGB color composition of the Fraser technique (**Figure 26**).

5.5 Lineament extraction

In geological structures, surface lineaments are often generated by deeply seated processes; such is often the case of faults and some mineral deposits. From satellite images, one can extract surface lineaments that may be associated with a given problem. In the present case, we can explore the association between surface lineaments and the hydrothermal alterations already mapped. To this end, we used Band 6 of the cropped Landsat image (**Figure 2**) and applied the four directional filters: N-S, E-W, NE-SW, and NW-SE, which represent the Freeman Code of Eight Directions (FCCE) [21]. The traces of the obtained lineaments are superposed to

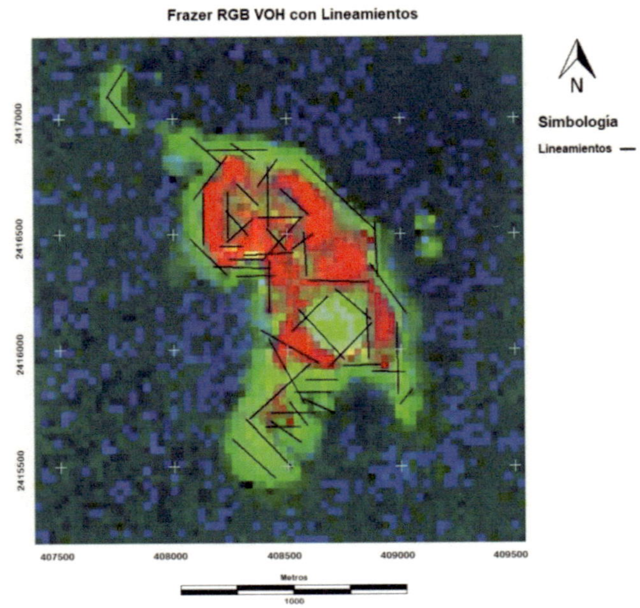

Figure 25.
RGB color composition image for the Fraser technique: vegetation (red), oxides (green), and hydroxyls (blue) including a set of lineaments (black lines) obtained from band 6 of the image, by means of directional filters.

Type	Number of pixels	Class
Ocean	3298	2, 4, 6
Vegetation	—	—
Oxides	—	—
Hydroxyls	963	7
Vegetation & oxides	197	5
Vegetation & hydroxyls	251	3
Oxides & hydroxyls	547	1, 8

Table 12.
*The K-means algorithm is applied to the color composite image of the Fraser technique (***Figure 25***).*

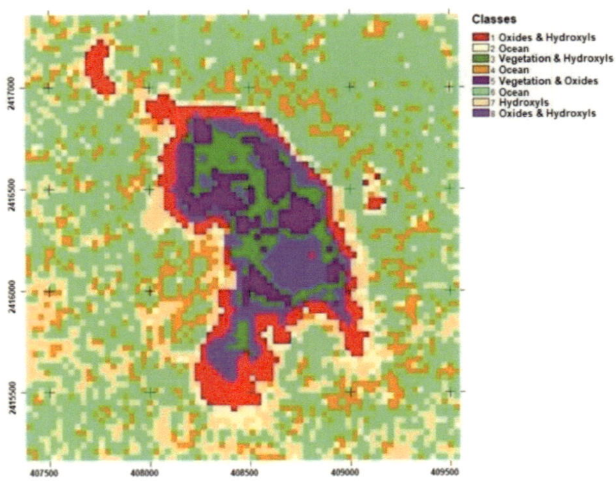

Figure 26.
*The K-means algorithm is applied to the RGB color composition image of the Fraser technique (***Figure 25***), with the indicated classes.* **Table 12** *shows the number of pixels identified in each class.*

the RGB color composition of the Fraser technique in **Figure 25**. A tendency is observed for the lineaments to follow a predominantly NW-SE direction, aligned with the main body of the island and with the orientation of the underlying laccolith [2], as well as with the main alignment of reported explosion craters [1]. The residual Bouguer anomaly of the island [3] also follows this direction suggesting that regional tectonic mechanisms favor this orientation. A more comprehensive discussion of the relation between lineaments and geophysical properties is beyond the scope of this study.

6. Conclusions

In summary, the comparison of these classifications: band ratio, PCA, Crosta, and Fraser techniques illustrates the difficulties and variants inherent in these classification methods. No single procedure appears to optimize all the parameters of interest. This set exemplifies potential variants in field explorations and shows the path for choosing the best alternative, or combination thereof, of interpretation for the problem at hand. For instance, the Fraser technique succeeded in identifying basalt flows covered by vegetation in the north half of the island but failed to locate the submerged basalts that were identified with the Crosta technique. In turn, the latter failed to identify the basalts covered by vegetation. Hydrothermal alterations are identified with different degrees of precision, indicating that if such is the target, one should evaluate various alternatives to optimize results. The analysis of the lineaments in the island, easily obtained by remote sensing, is associated with surface geological features and deep geophysical properties. They should be incorporated as frequently as possible into this type of studies.

The presence of hydrothermal activity in Isla Isabel extends almost fully throughout the island. Detection of such an activity under vegetated areas shows now that the northern half of the island is also the site of extended alterations. The possibility of basalt flooding in this region is high, and it should be the subject of detailed mapping. The high heat flow in the area, the existence of a laccolith under Isla Isabel, the extensive phreatomagmatic activity, and the presence of hydrothermal alterations in the island point to the relevance of this region as a potential geothermal source of major importance.

Acknowledgements

We acknowledge the expedient processing of permits to access the National Park by J. Castrejón, Director of Parque Nacional Isla Isabel. R. Lozano performed the geochemistry analyses in the X-ray Fluorescence Laboratory, his contribution is greatly appreciated.

Author details

Román Alvarez[1*] and Gerardo Figueroa[2]

1 Institute of Applied Mathematics (IIMAS), National University of Mexico (UNAM), Mexico City, Mexico

2 Earth Sciences Department, National University of Mexico (UNAM), Mexico City, Mexico

*Address all correspondence to: roman.alvarez@iimas.unam.mx

References

[1] Housh TB, Aranda-Gómez JJ, Luhr JF. Isla Isabel (Nayarit, México): Quaternary basalts with mantle xenoliths erupted in the mouth of the Gulf of California. Journal of Volcanology and Geothermal Research. 2009;**197**:85-107. DOI: 10.1016/j.jvolgeores.2009.06.011

[2] Alvarez R, Camargo FC, Yutsis V, Arzate J. A volcanic centre in Mexico's Pacific continental shelf. In: Németh K, Carrasco-Núñez G, Aranda-Gómez JJ, Smith IEM, editors. Monogenetic Volcanism. London. Special Publications 446: Geological Society; 2017. DOI: 10.1144/SP446.12

[3] Alvarez R, Camargo FC, Yutsis VV. Geophysical modelling of Isla Isabel: A volcanic island on the Mexican continental margin. In: Németh K, Carrasco-Núñez G, Aranda-Gómez JJ, Smith IEM, editors. Monogenetic Volcanism. London. Special Publications 446: Geological Society; 2017. DOI: 10.1144/SP446.13

[4] Reyes-Flores R. Petroleum exploration in the Pacific continental platform (states of Nayarit and Sinaloa). Ingeniería Petrolera. 1983:5-20. In Spanish

[5] Clark RN. USGS Spectroscopy Lab. https://speclab.cr.usgs.gov/

[6] Remote Sensing & GIS. Elementary Atmospheric Correction for Landsat Images: Haze Correction via Histogram Minimum. 2015. https://scientiaplusconscientia.worldpress.com

[7] Huang CP, Lavenburg G. Impacts of Bird Droppings and Deicing Salts on Highway Structures: Monitoring, Diagnosis, Prevention. Newark, Delaware: Civil and Environmental Engineering, University of Delaware; 2011

[8] Jensen JR. Remote Sensing of Environment. 2nd ed. London: Pearson Prentice Hall; 2007

[9] Available from: https://www.scribd.com/doc/162902456/Igpet-Manual

[10] Le Bas MJ, Le Maitre RW, Streckeisen A, Zanettin B. A chemical classification of volcanic rocks based on the total alkali silica diagram. Journal of Petrology. 1986;**27**:745-750

[11] Irvine TN, Baragar WR. A guide to the chemical classification of the common volcanic rocks. Canadian Journal of Earth Sciences. 1971;**8**:523-548

[12] Zanter K. Landsat 8 Data User Handbook. Sioux Falls, South Dakota: Department of the Interior U.S. Geological Suvey; 2016

[13] Intel Corporation. Open Source Computer Vision Library Reference Manual (PDF). 2001

[14] Ruiz-Armenta JR, Prol-Ledesma RM. Image procesing techniques for the exploration of ore minerals of hydrothermal origin. Física de la Tierra. 1995;7:105-137. In Spanish

[15] de la Vega-Márquez TF, Prol-Ledesma RM, Orozco G. Hydrothermal alteration and main structures mapping using TM images in La primavera geothermal field (Mexico). Geofísica Internacional. 2001;**40**:147-162

[16] Hartigan JA, Wong MA. Algorithm AS 136: A K-means clustering algorithm. Journal of the Royal Statistical Society: Series C (Applied Statistics). 1979;**28**(1):100-108. DOI: 10-2307/234683

[17] Abdi H, Williams LJ. Principal component analysis. Wiley Interdisciplinary Reviews:

Computational Statistics. 2010;**2**(4):433-459. DOI: 10.1002/wics.101

[18] IDRISI. https://clarklabs.org/terrset/idrisi-gis

[19] Lei L, Da-Fang Z, Jun Z, Dong-Sheng Q. Alteration mineral mapping using masking and Crosta technique for mineral exploration in mid-vegetated areas: A case study in Areletuobie, Xinjiang (China). International Journal of Remote Sensing. 2011;**32**(7):1931-1944. DOI: 10.1080/01431161003639678

[20] Fraser SJ. Discrimination and identification of ferric oxides using satellite thematic mapper data: A newman case study. International Journal of Remote Sensing. 1991;**12**:635-641

[21] Freeman H. On the encoding of arbitrary geometric configurations. IRE Transactions on Electronic Computers. 1961;**EC-10**:260-268

Chapter 3

Rock Physics Interpretation of Tomographic Solutions for Geothermal Reservoir Properties

Lawrence Hutchings, Brian Bonner, Seth Saltiel, Steve Jarpe and Mariel Nelson

Abstract

We present the basic theory of rock physics, laboratory studies, and field observations that can be applied to interpreting tomographic solutions from microearthquakes for reservoir properties. These properties include location of fluids, fractures, porosity, and permeability. We obtain 3D tomographic solutions for isotropic velocity (*Vp* and *Vs*) and attenuation (*Qp* and *Qs*), and derived solutions for elastic moduli (lambda, bulk and Young's), and Poisson's ratio. We show that a relatively high density of recording stations, about one each km^2, and relatively few microearthquakes (~300) can provide solutions at sufficient resolution and in a relatively short amount of time to be useful for reservoir exploration and management. We provide a case study for The Geysers, California, USA.

Keywords: tomography, reservoir modeling, microearthquakes, rock physics, The Geysers

1. Introduction

Microearthquakes with magnitude (M < ~3) occur naturally or due to fluid injection and production in geothermal, CO_2 sequestration, hydrocarbon, and natural gas reservoirs. In geological environments where sufficient numbers occur, they are a source of energy that provides information about reservoirs not often achieved by other methods. Recordings of microearthquakes can provide 3D tomographic solutions of isotropic velocity (*Vp* and *Vs*) and attenuation quality factors (*Qp* and *Qs*). These results can be used to also obtain solutions for Poisson's ratio, lambda, and Young's and bulk moduli. Furthermore, analysis of the microearthquakes provides locations, moment tensors and stress drops. These results can be combined with basic source theory and rock physics principles to interpret reservoir properties including regions of fluids, fractures, porosity, and permeability. We show that a relatively high density of recording stations, about one each per km^2, and relatively few earthquakes (~300) can provide solutions at sufficient resolution and in a relatively short amount of time to be useful for exploration and reservoir management. We provide a case study for The Geysers, California, USA.

IntechOpen

We do not attempt to obtain absolute values from tomography results, as this would require an excessively large number of events. We are interested in anomalies—changes in tomography results for areas with the same statistical resolution, depth, and comparable geology. Anomalies can indicate alterations and possible reservoir conditions. We make extensive use of automation to process the large amount of data collected [1, 2] and a tomographic inversion code designed to work with the output of the automated data processing (SimulCR; [3]).

2. Rock physics

Gassmann [4] first addressed rock physics by defining theoretical rocks based upon fractions of a sample of solid, liquid and gas. Later, more sophisticated approaches (effective medium theories: EMTs) represented rocks by additionally including pore shape (including fractures) and more quantitative descriptions (summarized by Berryman [5, 6]). These theories assume that response of a fractured or porous rock are heterogeneous on a macro-scale, but can be reproduced in a homogeneous rock that is equivalent to the former in the regime of static deformation. These additions better define the relationship among elastic parameters and material properties and therefore how they influence the propagation of seismic waves from microearthquakes. Of course, here and in subsequent discussions, we refer to general results, which sometimes vary.

Rock physics has been used to interpret recordings of active seismic sources primarily for oil and gas studies [6, 7]. Additionally, several authors have previously interpreted microearthquake studies with rock physics [8–14]. The effectiveness of seismic measurements to obtain reservoir properties has been successfully demonstrated for pore fluid pressures in country rock from injected CO_2 [15, 16] and fracture densities in underground repositories [17]. Authors have also used effective medium theory to further interpret tomography [1, 18–20]. Here we attempt to utilize a more comprehensive relationship between microearthquake recordings and material and fluid properties in a geothermal environment. The basic theory of rock physics, laboratory studies, and field observations that can be applied to interpreting tomographic solutions from microearthquakes for geothermal reservoir properties is outlined.

2.1 Seismic velocities

Hooke's Law shows that the strain field resulting from a generalized stress can be separated into a volume change with amplitude λ and a shape change with amplitude μ:

$$s_{ij} = \lambda \epsilon_{ii} \partial_{ij} + 2\mu \epsilon_{ij} \tag{1}$$

where s_{ij} is stress and ϵ_{ii} is strain.

The solution to the wave equation in terms of λ and μ specifies field observables V_p and V_s:

$$v_p = \sqrt{\frac{\lambda + 2\mu}{\rho}} = \sqrt{\frac{K + \frac{4}{3}\mu}{\rho}} \tag{2}$$

$$v_s = \sqrt{\frac{\mu}{\rho}} \tag{3}$$

where ρ is bulk density of the material and K is bulk modulus (below).

2.2 Elastic parameters

The elastic parameters used in this discussion can be described in terms of seismic velocity and density [7]. These solutions include density. We provide a density value of the aggregate state of the country rock with depth. Density can be removed by division from the solution for shear velocity (Eq. (3)). The parameters are as follows:

Poisson's ratio σ; the ratio of compressional or tensional strain to strain in orthogonal directions.

$$\sigma = \frac{V_p^2 - 2V_s^2}{2\left(V_p^2 - V_s^2\right)}. \tag{4}$$

Poisson's ratio is a valuable tool because it can be calculated from velocities alone and is insensitive to density variations caused by lithology. Seismologists have traditionally used *Vp*/*Vs* and Poisson's ratio interchangeably. However, it is evident that there is not a linear relation between *Vp*/*Vs* and Poisson's ratio, but over the range of values generally observed in seismic data ($0.2 < \sigma < 0.3$) the relationship is essentially linear. Poisson's ratio can range as low as 0.1 for foam, concrete, and dry, gas-saturated sands and as high as 0.5 for a perfectly elastic material, such as rubber at low strain. Cork has a value of 0.0. Effective medium theories (EMT) suggest that Poisson's ratio tends to vary smoothly with rock microstructure and elasticities of porous materials show that different pore shapes produce characteristic values for Poisson's ratio [6].

Shear modulus (μ): the relation of a shear stress to a shear strain in the same direction.

$$\mu = v_s^2 \rho. \tag{5}$$

Lambda (λ): the ratio of compressional or tensional stress to strains in orthogonal directions:

$$\lambda = \rho\left(V_p^2 - 2V_s^2\right). \tag{6}$$

Lambda is the off-diagonal component of the isotropic stiffness tensor in the absence of shearing effects and is referred to as incompressibility [5]. It is independent of μ. In the Gassmann model, λ is elastically dependent on fluid properties, while μ is not [5].

Young's modulus (E): the relation between the stress applied and the resulting strain in the same direction:

$$E = \rho V_s^2 \left\{ \frac{3R^2 - 4}{R^2 - 1} \right\} \tag{7}$$

with

$$R = \frac{V_p}{V_s} \tag{8}$$

Little research has been done on the relation of Young's modulus to reservoir properties, but we execute an analysis in this study to see if any effects can be identified.

Bulk modulus (K): a measure of how compressible a material is. It relates the volume's change in shape resulting from triaxial or hydrostatic stress:

$$K = V\frac{\partial P}{\partial V} = \rho\left\{V_p^2 - \frac{4V_s^2}{3}\right\}. \quad (9)$$

where V is volume and P is pressure. So, K approaches 0.0 for a fully rigid body. Hereafter lambda, bulk modulus, Young's modulus, and Poisson's ratio are referred to as λ, K, E, and σ, respectively.

2.3 Attenuation

Attenuation is the loss of energy with wave propagation, and Q is the quality quotient that describes the amount of attenuation (Aki and Richards, p. 220). Q has the reciprocal effect of attenuation, i.e., high Q is low attenuation and low Q is high attenuation. Attenuation is generally assumed to be due to inter-crack motion or fluid flow between pores, and is generally called intrinsic attenuation. Extrinsic attenuation is apparent attenuation when seismic energy is scattered due to small fractures. Energy is not actually lost, but a wavefront pulse of a propagating arrival broadens with distance, as would be observed with actual attenuation. Menke [21] fit observations for dry competent rock, and explains how attenuation decreases with depth due to crack closure and stiffening. Tokzoz and Johnson [22] also say this explains why many laboratory studies show attenuation to be frequency independent. In our analysis we do not distinguish pulse broadening due to intrinsic or extrinsic Q. This is determined by comparisons to other attributes. We also assume frequency independent Q.

2.4 Cracks, fractures and faults

Several mechanisms exist for the creation and destruction of permeability at depth; all of these involve cracks, fractures or faults. Cracks are presumed to be associated with weak grain boundaries, fractures are at the scale of multiple cracks, and faults are dislocations from earthquakes. In the most general case, the nucleation and propagation of cracks may increase the connectivity between cracks and fractures, and thus permeability [23]. Fractures alone can be conduits of permeability, as observed in the Salton Sea [19]. Faults are often conduits for fluid flow as well and thus also affect permeability. Faults are considered tectonic in nature and are often aligned with regional stress patterns.

Accurate locations of microearthquakes can often denote permeable zones while their moment tensors can identify the orientation and type of fractures being formed. Guilham et al. [24], Johnson [25] and Julian [26] interpreted moment tensors to obtain focal mechanism solutions that indicate the existence of isotropic and deviatoric dislocation events at The Geysers.

The rupture process and resulting permeability from microearthquakes may not be identical at all scales. Microearthquakes that rupture an entire crack or fracture may have an end effect, i.e., deformation at the end of a fracture to accommodate slip. Johnson [27] theorizes that this gives rise to orthogonal tensile crack opening at the end of cracks or fractures. Microearthquakes associated with faults may rupture only part of a larger feature created by previous earthquakes, thus not create tensile cracks.

Rocks with parallel crack or fracture systems will cause anisotropic wave propagation and shear-wave splitting will occur. In geothermal environments, shear-wave velocity anisotropy has been observed to as high as 10% [28, 29]. Several authors attempt to address anisotropy as a tomography problem in order to identify where it occurs [30, 31]. Rose diagrams [28, 32] only map observed shear-wave splitting and do not identify locations of its occurrence.

Cracks and fractures (C&F) decrease μ without significantly reducing ρ and thus decrease both *Vp* and *Vs*. In the presence of C&F a geologic material is expected to have low *Vp* and *Vs* near the surface due to low μ and generally increase monotonically with depth due to the closure of cracks and fractures as pressure increases from the lithostatic load [33, 34]. Basement rocks at lithostatic pressures consistent with depths of greater than 3 km can be expected to have very low permeability (< 10–3 μ Darcy) due to closure of fractures [35]. At these depths permeability can be due to micro-fractures. Once C&F close, velocity will no longer increase with depth. O'Connell and Budiansky [36] relate a C&F density parameter to the effective Poisson's ratio, which is a direct reflection of their effect on *Vp* and *Vs*. λ, B, and E will also increase with depth due to compliance from open C&F.

In hydrothermal environments, grain-scale geochemical reactions can cement micro-fractures and stiffen the rock matrix. As a result, their relative effect on velocity will be reduced. This is observed in laboratory measurements of velocities in recovered core from The Geysers [33]. This suggests that much of the observed depth dependence of *Vp* and *Vs* in The Geysers reservoir rocks is due to closure of larger-scale C&F. Large permeable fractures might then be identified by regions of high velocity gradients in field data with a high density of compliant fractures. This may be particularly true in geothermal areas where healed micro-fractures will contribute less to observed gradients [33, 37].

Attenuation is also significantly affected by C&F. C&F increase extrinsic attenuation and reduce Qp and Qs. The lithostatic load tends to decrease extrinsic attenuation with depth due to closure of C&F. Extrinsic attenuation is potentially different for *P*-wave propagation and thus Qp than for *S*-wave propagation and Qs if C&F are aligned, since particle motion is transverse.

2.5 Effects of fluids and steam

Following Eqs. (2) and (3), the inclusion of fluids into either pores or fractured material increases density and decreases *Vp* and *Vs*. Fluids also increase the bulk modulus, which increases *Vp*. In low porosity rocks, increased bulk modulus generally dominates the increase in density, whereas the reverse is true for high porosity rocks. The shear modulus (μ) is independent of fluids in the absence of geochemical effects and is determined by the porous rock matrix [6]. Therefore, density changes are the only effect that fluids have on *Vs*.

Injection of fluids can cause micro-fractures and/or microearthquakes due to thermal contraction or hydro-fracturing. Fluids also significantly affect attenuation. Partial saturation increases intrinsic attenuation. However, full saturation should lower intrinsic attenuation by inhibiting diffusion. Diffusion is also different for *P*-wave propagation and thus Qp than for *S*-wave propagation and Qs because particle motion is transverse. We note that Qs decreases and Qp increases with saturation (**Figures 5** and **6**).

Berryman et al. [8] emphasized that lambda and density contain information about saturation, while both combined with shear modulus contain information about porosity. Berryman et al. recast λ (Eq. (6)) as λ/μ, removing density $(\rho = \mu/v_s^2)$ and showing lambda's application in identifying the degree of saturation and type of saturation, i.e., the arrangement of fluids in a rock (inhomogeneous or homogeneous). They showed that for homogeneous saturation, λ/μ remains low for partial saturation and high for full saturation, and for inhomogeneous saturation there is a monotonic increase in λ/μ as saturation increases (equal at full saturation).

Typically, seismologists use high values of Poisson's ratio (or *Vp/Vs*) to indicate fluid saturation. There is a dramatic increase in Poisson's ratio with saturation by the replacement of vapor by water, which increases the effective bulk modulus, while *Vs* is unaffected by fluid content.

Laboratory work suggests that using a combination of attenuation and velocity can improve discrimination of pore fluid content. Winkler and Nur [38] measured moduli and attenuation for longitudinal and torsional modes in porous and cracked rock near 1 kHz, where intrinsic attenuation dominates extrinsic attenuation. Plotting the data in *Qp/Qs* and *Vp/Vs* coordinates separates dry, partially saturated, and fully saturated conditions. These data cannot be used quantitatively until more measurements are done with low frequency data. The strong trends in Winkler and Nur's data suggest, however, that field data should reflect similar effects. Thus, values above a slope of 1.0 for *Qp/Qs* versus *Vp/Vs* plot indicate saturation and lower values indicate drier conditions. Hutchings et al. [3] found that The Geysers' data supported Winkler and Nur's theory.

Gritto [39] theorized that when injected water contacts reservoir rock, heat is drawn from the reservoir rock until the water vaporizes. The resulting cooling and contraction of the rock generates tensile (mode I) cracks and subsequent micro-seismicity. Once all water is converted to steam, the rock remains at a stable temperature with no further seismicity and the reservoir has reached maximum steam concentration. Permeability can be measured by monitoring the spatial and temporal migration of the micro-seismic cloud associated with fluid injection [40]. The resulting cooling may also result in increased shear moduli.

2.6 Temperature and pressure effects

Rocks subjected to high temperatures and pressures undergo a transition from brittle to crystalline plastic behavior. The temperature of this transition ranges from around 300°C (quartz) to 400–450°C (feldspar) and also depends on pressure and strain rate [41, 42]. When stressed, some rocks may undergo cataclastic flow, which is characterized by ductile stress-strain behavior as well as cracking and frictional sliding.

The effect of temperature can oppose the effect of pressure [34]. In a dynamic situation, heating or cooling of fluids within pores can cause fractures and increase permeability while pressure can close fractures. Experiments where granite was cooled from high temperatures up to 646°C showed permeability increased up to a 1000 times over original values [43]. Darot et al. also found permeability decreased rapidly with confining pressure, being effectively zero for confining pressures over 30 MPa.

2.7 Summary: porosity, permeability, and saturation

There are seven primary interpretations of porosity, permeability and saturation from observable microearthquake data. Comparisons are made relative to normal geology at similar depths and temperatures, meaning geology that has a monotonic increase in velocity and *Q* as a function of depth, and saturation, porosity and temperature that is considered average for the geologic condition of the study area. The interpretations are as follows:

1. *Dry competent geology with low porosity* might be identified by increased *Vp* and *Vs* due to high shear modulus due to few C&F; high *Qp* and *Qs* due to lack of diffusion and little extrinsic attenuation due to few C&F; Poisson's ratio near 0.25; lambda, bulk and Young's modulus are relatively high due to incompressibility of stiff material. No fracturing or permeability is assumed.

2. *Dry competent geology with high porosity* might be identified by a slight increase in *Vs* due to lower density, and a decrease or no change is *Vp* due to low bulk modulus offsetting low density with *Qp* and *Qs* relatively unchanged. Poisson's ratio is slightly decreased due to higher Vs and lower Vp. Young's modulus and lambda are relatively low due to compressibility and dry conditions. No fracturing or permeability is assumed.

3. *Dry fractured geology* might be identified by lower *Qp* and *Qs* due to an increase in extrinsic attenuation and lower *Vp* and *Vs* due to a decrease in shear modulus. Poison's ratio, bulk modulus, Young's modulus and lambda are decreased due to increased compressibility. No permeability is assumed, and conditions are the same if the material is high or low porosity.

4. *Saturated competent geology with low porosity* might be identified by slightly higher *Qp* due to reduced intrinsic attenuation and *Qs* may be low due to saturation. A small decrease in *Vs* is expected due to an increase in density with slightly higher or unaltered *Vp* due to an increase in bulk modulus (possibly offset by increased density). Poisson's ratio and lambda will be higher, shear modulus unchanged, and Young's modulus slightly higher due to decreased compressibility. Partial saturation reduces these effects.

5. *Saturated competent geology with high porosity* might be identified by slightly higher *Qp* due to reduced intrinsic attenuation and *Qs* may be low due to saturation. A significant decrease in Vs is expected due to an increase in density and little change or increase in *Vp* due to an increase in density and a counter-effect of increased bulk modulus. Poisson's ratio and lambda will be significantly higher with shear modulus unchanged. Young's modulus will be higher. Permeability might be assumed. Partial saturation reduces these effects.

6. *Saturated fractured geology with high porosity* might be identified by significantly lower *Qp* and *Qs* due to an increase in both extrinsic and intrinsic attenuation; a significant decrease in *Vs* due to an increase in density and a reduction of the shear modulus due to fractures; and an increase in *Vp* due to the dominating effect of high bulk modulus (or lambda). The increase in density and decrease in shear modulus have a mitigating effect. Permeability is likely high due to intersecting and open fractures. Poisson's ratio is high. Bulk modulus, Young's modulus, and lambda will be higher due to saturation and incompressibility.

7. *Saturated fractured geology with low porosity* might be identified by lower *Qp* and *Qs* due to an increase in extrinsic, but little increase in intrinsic *Q*. Moderately low *Vs* is expected due to a decrease in shear modulus with little increase in density and lower or unaltered *Vp* is expected due to the competing effects of decreased shear modulus and increased bulk modulus. Permeability is likely high due to intersecting and significantly open fractures. Poisson's ratio is unchanged or moderately higher. Young's modulus and lambda are significantly higher due to saturation.

3. Case study: the Geysers

Hutchings et al. [3] analyzed a ~5.5 km^3 volume at the Northwest Geysers area (**Figure 1a**). Seven injection wells were in operation in this volume during the period of the study (**Figure 1b**). The known locations of induced events, the timing, rate, and location of water injection, and the location of geologic alterations at the bottom of injection wells provide a test of tomography and a rock physics approach to reservoir property analysis. The study is divided into two time periods, approximately 42 days before and 31 days after injection rates changed significantly at several of the wells. **Figure 1** shows microearthquake locations, recording station locations, and wells utilized in the study.

The study area is within a portion of The Geysers geothermal field that has a high temperature reservoir (HTR, with temperatures up to 400°C) that underlies a normal temperature reservoir (NTR, with temperatures about 260°C) shown in **Figure 2** [44]. The HTR extends downward from about ~2 km below mean sea level (bmsl) in the southwest to ~3 km bmsl in the northeast. The NTR reservoir extends between ~1.5 and ~2 km bmsl [44]. The rocks are composed of relatively permeable greywacke in the NTR and of low-permeability, thermally altered greywacke in the HTR.

We separated data into two recording time periods and performed tomography, the first from 1 September to 8 October 2011 and the second from 9 October to 11 November 2011. We refer to these as period 1 and period 2 in the text. Tomography was performed with 23 recording stations, about one station per 1.3 km^2, and with 378 and 369 earthquakes for the two time periods. The first time period is for 42 days before and the second is for 31 days after injection rates changed significantly at several of the wells.

Figures 3 and **4** show *Vs* and *Vp* tomography results for a cross section that passes through the bottom of wells WH34 and Prati-32 along with a cross section that passes through the bottom of wells Prati-9 and Prati-54. These are shown as AA′ and BB′ in **Figure 1**. Also shown is the resolution value for diagonal of the resolution matrix for the tomography. **Figure 3a** shows the thickening of the Greywacke seen in **Figure 2**. There is very little change in the *Vp* tomography

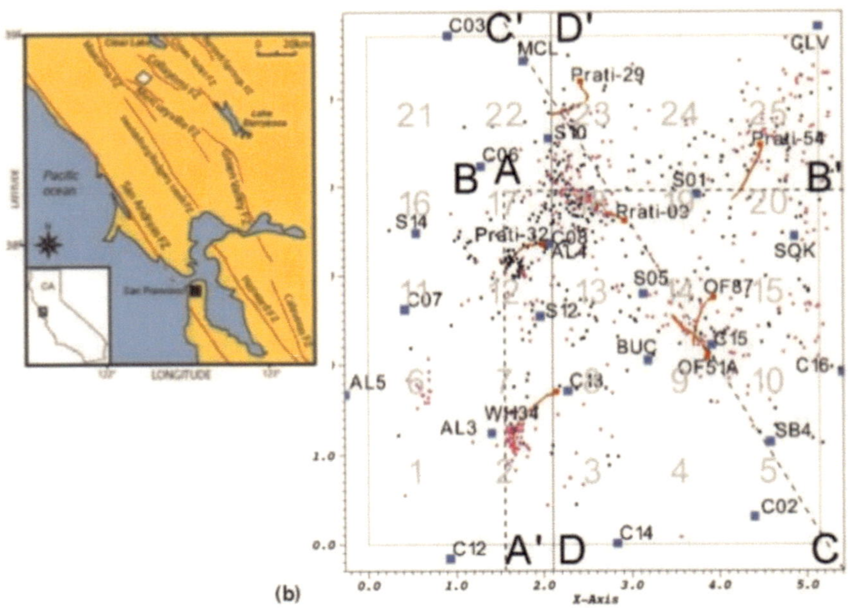

Figure 1.
(a) Regional setting and the location of the study area (white diamond) and locations of major faults. (b) Map view of study area—recording stations in blue squares, earthquakes from first period in magenta dots and those during the second period in black dots, wells in red (red square is top of well), and the cross section locations AA′, BB′, CC′, and DD′ discussed in text. The study area is rotated 39° clockwise, so CC′ is aligned almost due north.

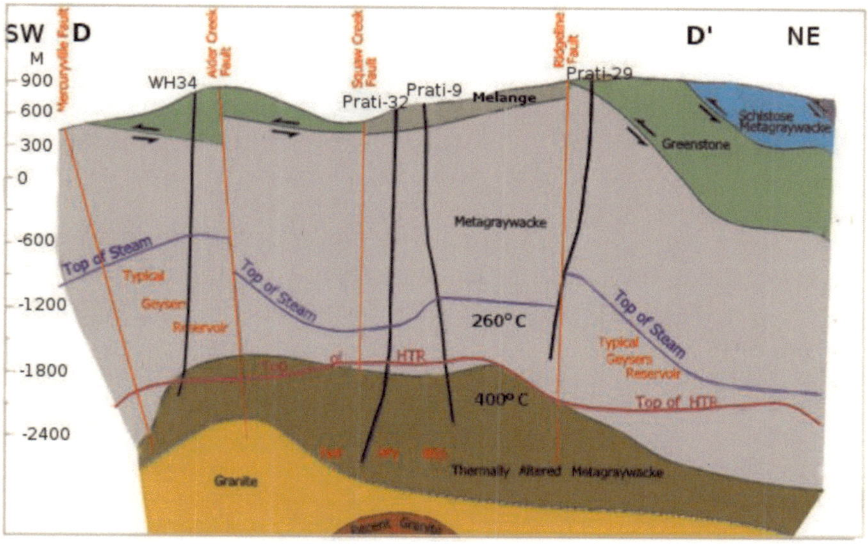

Figure 2.
Geologic cross section along DD′ in ***Figure 1b****, view is northwest; modified from [44].*

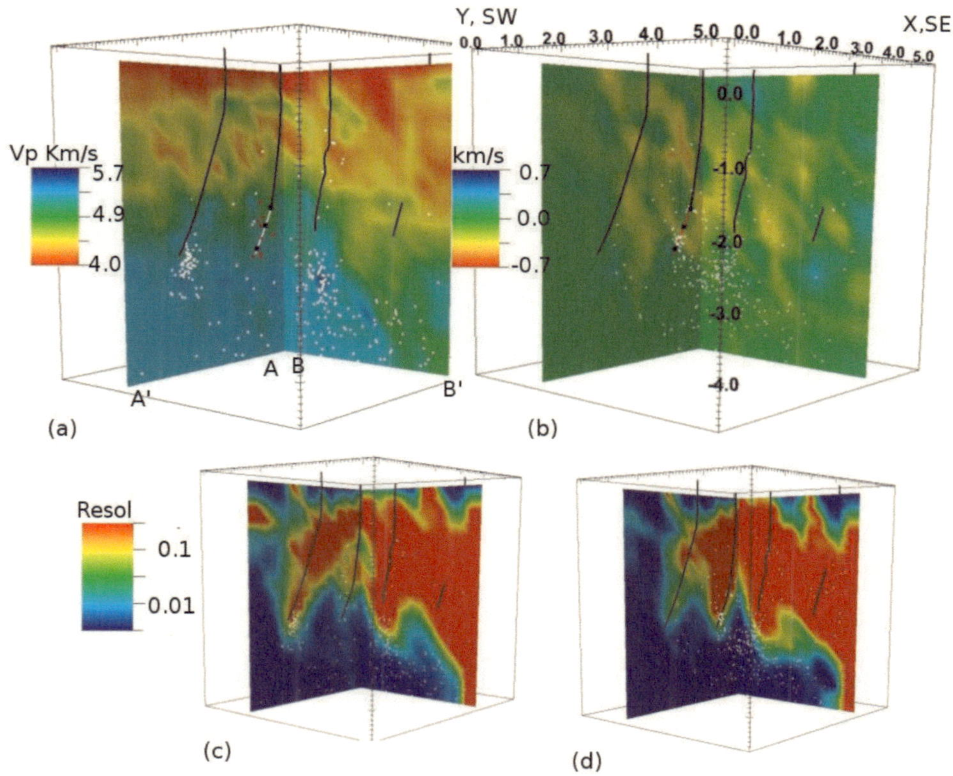

Figure 3.
(a) Vp tomography results for period 1 along cross sections AA′ and BB′ shown in ***Figure 1b****; (b) changes in Vp results for period 2, i.e., ΔVp; (c) resolution for period 1; (d) resolution for period 2.*

results for the second time period. The red dots at the bottom of Prati-32 are location of the first events that occurred after injection started (located with the tomography results from the first time period). The black squares are the steam entry points. That the first few events occur near the wells and near the

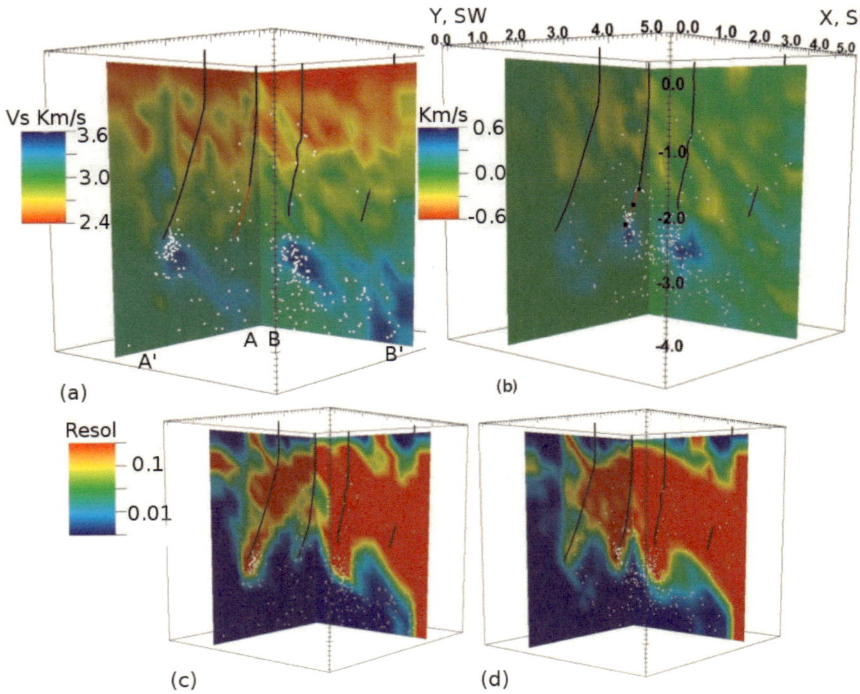

Figure 4.
(a) Vs tomography results for period 1 along cross sections AA′ and BB′ shown in ***Figure 1b****; (b) changes in Vs results for period 2, i.e., ΔVs; (c) resolution for period 1; (d) resolution for period 2.*

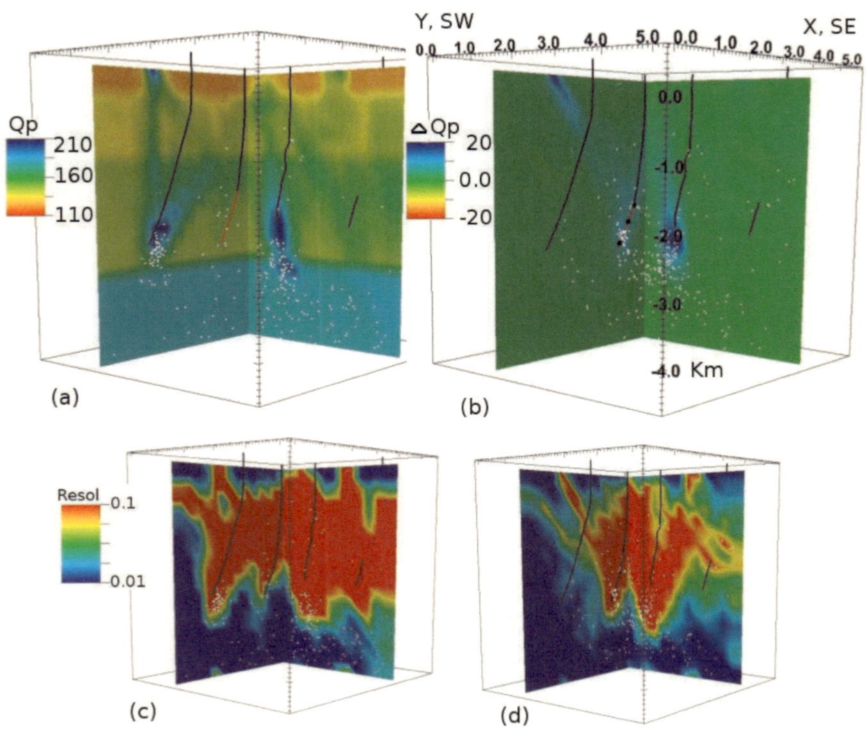

Figure 5.
(a) Q p tomography results for period 1 along cross sections AA' and BB' shown in ***Figure 1b****; (b) results for ΔQ p for period 2, Q p scale on left; (c) resolution for period 1; (d) resolution for period 2. Resolution scale is for diagonal of resolution matrix, where red is greater than 0.1 and blue is less than 0.01.*

steam entry points (and presumably the water release points) and validates the accuracy of the earthquake locations, and subsequently the tomography results.

From **Figure 4a**, there are relatively high *Vs* anomalies below WH34, Prati-9 and possibly Prati-54 (extreme lower right) with no anomaly below Prati-32. These anomalies indicate about a 20% increase in *Vs*. **Figure 4b** shows the change in tomography results for *Vs* during period 2. The anomaly below Prati-9 shows a further increase in *Vs* and a new anomaly appeared below Prati-32. The anomalies below WH34 and Prati-54 have remained unchanged. The high velocity anomalies below WH34, Prati-32 and Prati-9 extend from the bottom of the wells and are not randomly occurring; we conclude they are not artifacts. Seismicity is distributed around the wells and throughout the deeper portions of the volume (>−1.0 km). The resolution in the anomalous areas below the wells is in the range we consider acceptable for our purposes. The new anomaly below Prati-32 during period 2 demonstrates that temporal changes in reservoir properties can be observed over at least a month.

Figures 5 and **6** show *Qp* and *Qs* tomography results for the same cross sections as for **Figures 3** and **4**. The value for the diagonal of the resolution matrix is also shown. **Figures 5b** and **6b** show the changes in *Qp* and *Qs* for the second time period, respectively. It is apparent from comparing the background *Q* values obtained in the inversion for the two time periods that the tomography has only significantly changed around the well bottoms. We conclude that the anomalies at the bottom of wells WH34 and Prati-9 are significant. *Qp* increased and *Qs* diminished considerably. The *Q* anomalies envelop the base of the wells and are not

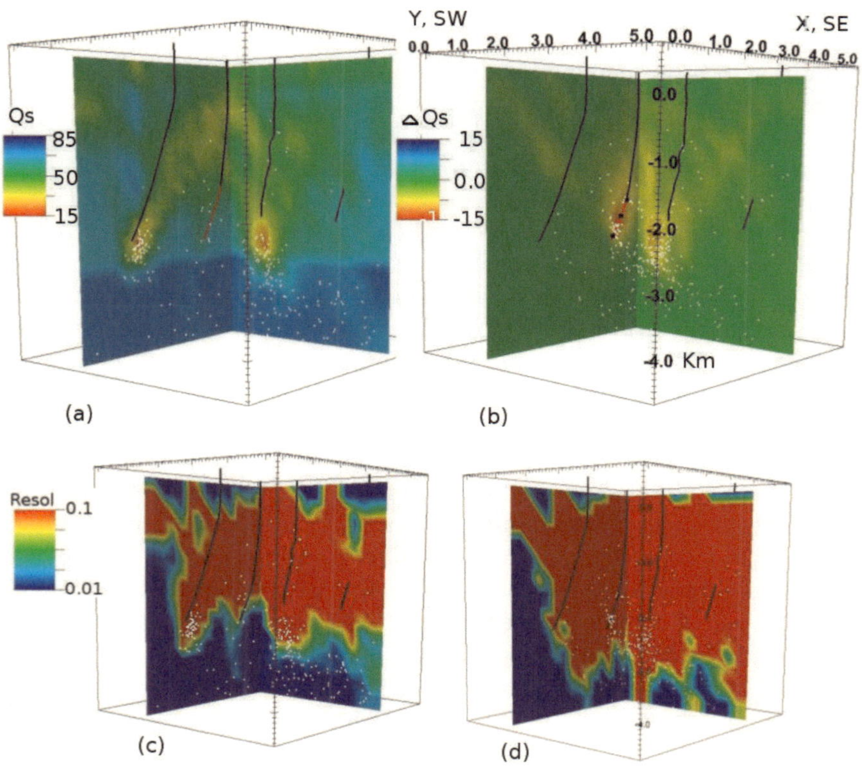

Figure 6.
(a) Qs tomography results for period 1 along cross sections AA' and BB' shown in ***Figure 1b****; (b) results for ΔQs for period 2, Qs scale on left; (c) resolution for period 1; (d) resolution for period 2. Resolution scale is for diagonal of resolution matrix, where red is greater than 0.1 and blue is less than 0.01.*

located exactly where the *Vp* and *Vs* anomalies are observed. No other anomalies are apparent in the results. The anomaly at the bottom of Prati-9 increased in size and the anomaly below WH34 remained unchanged. The background values remained unchanged for the two time periods, so we are fairly confident the anomalies are real and the change below Prati-9 is real.

3.1 Interpretation

Our primary interpretation is that the anomalies we observe are the result of cold water injected into hot material. The wells below which no alteration was observed, and no earthquakes, were Prati-29, Prati-54, OF87A, OF51A and Prati-54. These are located at shallower depths and at cooler temperatures than wells WH34, Prati-32, and Prati-9, where alterations were observed and which are within the hot deep zone. Possible alterations below Prati-54 and OF51A exist at depths far enough below the wells to be located in the hot zone. These observations support our hypothesis that the alterations we observe are a result of cold water injected into very hot geology, which also causes earthquakes. Further, injected water has percolated down to very hot geology. Thus, water responding to gravity reaches as deep as 1.5 km below the wells (and possibly 4.5 km below Prati-54), causing the anomalies.

We further hypothesize that injection cools rock near the well bottoms, which stays cool and saturated as injection continues. Cooling near the well bottoms generates tensile cracks and subsequently micro-seismicity. As water turns to steam, a pressure front triggers more earthquakes away from the wells. However, there is not a concentration of seismicity in the high *Vs* anomalies, but throughout the deep zone, suggesting that the hot deep geology is not ductile enough to be aseismic.

In addition to shear modulus, lambda also decreases (not shown), offsetting the increase in shear modulus so the change in *Vp* is not pronounced. Density does not increase because the fluid turns to steam. Poisson's ratio gets extremely low (not shown), mostly due to the significant increase in *Vs* with little change in *Vp*. Young's modulus is high and bulk modulus is low (not shown), which we interpret as pores filled with steam, but in other portions of the surrounding deep zone, Young's modulus is high and bulk modulus is normal, which we attribute to normal rock properties. In Eq. (6), lambda decreases in value by $-Vs^2$, so a low value of lambda is not surprising. In Eq. (9), bulk modulus is similarly reduced by a factor proportional to $-Vs^2$. In Eqs. (7) and (8), Young's modulus is proportional to $+Vs^2$, so its increase is also not surprising.

There are high *Qp* and low *Qs* anomalies at the bottom of wells WH34, Prati-32 and Prati-9. The cold water in hot material causes fracturing and a significant increase in seismicity in addition to saturation (initially, before the water turns to steam). If a region is partially saturated, one would expect both intrinsic and extrinsic *Q* to be low, which could account for the low *Qs* anomaly. But the high *Qp* anomaly poses several issues. If material is fully saturated, intrinsic *Q* would be high, as there would be no movement between pores. However, *Qs* would still reflect the fractures because *Vs* is relatively unchanged by the presence of water. *Qp* might be high due to high intrinsic Q and high bulk modulus, so perhaps fractures are not reflected in *Qp*, meaning extrinsic *Qp* would also be high. Furthermore, the particle motion of shear waves is orthogonal to the compressional motion of the *P*-waves, alignment of fractures may also have an effect. Fractures parallel to *P*-wave particle motion would not alter *Q* and fractures perpendicular to shear-wave particle motion perpendicular to fractures would alter *Qp*. Since our tomography does not differentiate between the geometry of attenuation types, we cannot say for sure which is occurring. There are some studies that support the difference between Qp and Qs under saturation conditions.

Tokzoz et al. [45] and Johnston and Tokzoz [46] both show that under dry conditions $Qp \sim Qs$, but under full saturation $Qs < Qp$, agreeing with our observations. This occurs under pressure but at ultrasonic frequencies. DeVilbiss-Munoz [47] shows that Qp and Qs increases significantly as water turns to steam and Mavkov and Nur [48] show Qs increases relative to Qp as saturation increases. Neither of these are consistent with our interpretation.

3.2 Reservoir analysis

We examine observations of normal temperature reservoir depths in the greywacke, from ~1 to 2.5 km bmsl. The top of the steam/liquid boundary is at about 1 km bmsl and deepens to the southeast (**Figure 2**). Lowenstern and Janik [49] point out that the northwest portion of The Geysers has little condensed liquid and contains primarily steam-filled pores. Wells Prati-29 and OF87A-11 are shallower than the other wells in this study and were drilled into the greywacke. OF51A-11 and Prati-54 are deeper, but are located in the eastern portion of the field where the greywacke deepens, so their temperature is comparable to the shallower wells and is included in this discussion. Interestingly, no anomalies in our eight attributes are observed at the bottom of these wells. Temperatures are near 240°C and are apparently too low to induce the effects seen at the other wells drilled into the thermally altered greywacke with temperatures near 400°C or higher.

To identify anomalies at reservoir depths, we used the tomography results for the second time period, which used the results from the first time period as a starting model. Results in the depth range of interest here did not change significantly for the second time period (**Figures 3–6**). We examined values of the eight parameters for voxels that are 1 × 1 km laterally and 0.75 km in depth throughout the study volume. Each voxel is identified by whether the of majority of the volume visually shows high, low or no anomaly. Values are relative to what would be expected for normal geology at comparable depth and resolution. These observations are put into rows of **Table 1**, where one row is for each voxel; there are fifty rows for our approximately 5 × 5 km study area.

Figure 1 shows the surface projection of the location of the volumes, two volumes for each ID number. The ID numbers are listed in the first row of **Table 1**. The second column identifies the center of the voxel in kilometers relative to the origin for X and Y. Regarding the eight attributes, "0" indicates no anomaly, "+" a positive and "–" a negative anomaly. Generally, an anomaly is identified as being at least plus or minus 20% of average values at the same depth and geology. The final column provides a rock physics interpretation based on the eight conditions described in the introduction, which are derived from the basic principles outlined in Section 2. Zero change across the row indicates standard reservoir conditions, which is likely greywacke with steam-filled pores. It is assumed that there is not strict compliance with the eight descriptions in Section 2.8—for example, a voxel with a "0" value in the table may still show a slight change.

The anomalies identified in **Table 1** can be seen in some of the cross sections. **Figure 1** cross section CC′ is aligned almost north-south, so cross section AA′ represents the western portion and BB′ represents the eastern portion of the study volume; the upper part of CC′ represents the northern part and the bottom part of CC′ represents the southern part through the middle of the study area.

ID	*X*, *Y*	*Vp*	*Vs*	*Qp*	*Qs*	Poiss. ratio	Lam.	Bulk	Young's	Interpretation
1	0.6, 0.7	0	0	0	0	0	0	0	0	Standard reservoir
		0	0	0	0	0	+	+	0	(4) Saturated competent geology with low porosity
2	1.6, 0.7	0	0	0	0	0	0	0	0	Standard reservoir
		0	0	0	0	+	+	+	0	(4) Saturated competent geology with low porosity
3	2.6, 0.7	–	0	0	0	0	–	0	0	(2) Dry competent geology with high porosity
		0	0	0	0	0	0	+	0	Standard reservoir
4	3.6, 0.7	0	0	0	0	0	0	0	0	Standard reservoir
		0	0	0	0	0	0	0	0	Standard reservoir
5	4.6, 0.7	0	0	0	0	0	0	0	0	Standard reservoir
		0	0	0	0	0	0	0	0	Standard reservoir
6	0.6, 1.7	0	0	0	0	0	0	0	0	Standard reservoir
		0	0	0	0	0	0	+	0	Standard reservoir
7	1.6, 1.7	0	+	0	+	0	0	0	0	(1) Dry competent geology with low porosity
		0	0	+	0	0	+	+	0	(4) Saturated competent geology with low porosity
8	2.6, 1.7	0	0	0	0	0	–	0	0	Standard reservoir
		0	0	0	0	0	0	0	0	Standard reservoir
9	3.6, 1.7	0	0	0	0	0	0	0	0	Standard reservoir
		0	0	0	0	0	0	+	0	Standard reservoir
10	4.6, 1.7	0	0	0	0	0	–	0	0	Standard reservoir
		0	0	0	0	0	–	0	0	Standard reservoir
11	0.6, 2.7	0	0	0	0	0	0	0	0	Standard reservoir
		0	0	0	0	0	+	+	0	(4) Saturated competent geology with low porosity
12	1.6, 2.7	0	0	+	0	0	0	0	0	Standard reservoir
		0	0	0	0	0	0	0	0	Standard reservoir
13	2.6, 2.7	0	0	0	0	0	0	0	0	Standard reservoir
		0	0	0	0	0	0	0	0	Standard reservoir
14	3.6, 2.7	0	0	0	0	0	–	0	0	Standard reservoir
		0	0	0	0	0	0	0	0	Standard reservoir
15	4.6, 2.7	–	0	0	0	0	0	0	0	Standard reservoir
		0	0	0	0	0	–	0	0	Standard reservoir
16	0.6, 3.7	0	0	0	0	0	0	0	0	Standard reservoir
		0	0	0	0	0	+	+	0	(4) Saturated competent geology with low porosity
17	1.6, 3.7	0	0	0	0	0	0	0	0	Standard reservoir
		0	0	0	0	0	+	+	0	(4) Saturated competent geology with low porosity

ID	X, Y	Vp	Vs	Qp	Qs	Poiss. ratio	Lam.	Bulk	Young's	Interpretation
18	2.6, 3.7	0	0	+	0	0	0	0	0	Standard reservoir
		+	0	+	–	+	+	+	0	(5) Saturated competent geology with high porosity
19	3.6, 3.7	–	0	0	0	–	–	0	0	(2) Dry competent geology with high porosity
		0	0	0	0	0	0	0	0	Standard reservoir
20	4.6, 3.7	0	0	0	0	0	0	0	0	Standard reservoir
		0	0	0	0	0	–	0	0	Standard reservoir
21	0.6, 4.7	0	0	0	0	0	0	0	0	Standard reservoir
		0	0	0	0	0	+	+	0	(4) Saturated competent geology with low porosity
22	1.6, 4.7	0	0	0	0	0	0	0	0	Standard reservoir
		0	0	0	0	0	+	+	0	(4) Saturated competent geology with low porosity
23	2.6, 4.7	0	0	0	0	0	0	0	0	Standard reservoir
		+	0	0	0	0	+	+	0	(4) Saturated competent geology with low porosity
24	3.6, 4.7	–	0	0	0	0	–	0	0	(2) Dry competent geology with high porosity
		0	0	0	0	0	–	0	0	Standard reservoir
25	4.6, 4.7	–	0	0	0	–	–	0	0	(2) Dry competent geology with high porosity
		0	0	0	0	–	–	0	0	(2) Dry competent geology with high porosity

Table 1.
Location of quadrants where tomography attributes are examined.

4. Discussion and conclusions

We analyze eight attributes tomographic images obtained from tomographic images: isotropic velocity (*Vp* and *Vs*), attenuation (*Qp* and *Qs*), and derived elastic moduli (lambda, bulk and Young's) and Poisson's ratio, in addition to earthquake locations. The known locations of induced events, the timing, rate, and location of water injection, and the location of geologic alterations at the bottom of injection wells provide a test of this rock physics approach to reservoir property analysis. We outline rock physics principles that can be used to interpret reservoir properties from these observations. We demonstrate that using a relatively high density of stations and examining anomalies, we obtain results in a shorter time period, with higher accuracy, and with fewer earthquakes than is typical for reservoir studies. We also apply a systematic rock physics evaluation of 50 1 km^3 volumes at reservoir depths and demonstrate the ability to identify reservoir properties. In the deeper portion of the volume (near the well bases, below the existing reservoir), seven of the eight attributes show significant effects of cold water injected into hot material and variations over a two-month time span. The results also suggest water is penetrating as deep as 1.5 km and possibly 4.5 km below the wells, even though temperatures reach at least 400°C in the country rock. This causes an increase in shear modulus due to cooling, however, due to the temperature, the water quickly turns to steam.

We consider explanations for the relatively high *Vs* estimates in the hot deep zone exhibited in **Figure 3**. Noted earlier, these anomalous regions occur at depths comparable to or deeper than the well termination depths within the HTR (**Figure 2**). This is where the natural reservoir temperatures increase from a relatively homogenous 240°C in the NTR (~900–1800 m bmsl) to at least 400°C, measured at the base of Prati-32 (2672 m bmsl), the deepest wellbore [44]. The pockets of high *Vs* observed in period 1 appear to spatially correspond with wells where water was actively being injected. Furthermore, during period 2, these anomalies intensified only below well bores in which the injection rate increased substantially relative to period 1 (Prati-32, Prati-9, **Figure 3**), including the unambiguous appearance of a new anomaly below Prati-32 associated with the injection of water into a previously undisturbed region of the HTR. The enhanced geothermal system (EGS) demonstration at Prati-32 is therefore an exemplary scenario for considering likely mechanisms to account for the observed evolution of higher *Vs* zones in the HTR, requiring changes in the physical properties of the reservoir material in the range of 30% decrease in bulk density or a 45% increase in shear modulus.

Author details

Lawrence Hutchings[1*], Brian Bonner[1], Seth Saltiel[2,3], Steve Jarpe[4] and Mariel Nelson[2]

1 Lawrence Berkeley National Laboratory, California, United States

2 Department of Earth and Planetary Sciences, University of California, Berkeley, United States

3 Department of Geophysics, University of Chile, Santiago, Chile

4 Jarpe Data Solutions, Inc., Berkeley, California, United States

*Address all correspondence to: ljhutchings@lbl.gov

References

[1] Hutchings L, Bonner B, Jarpe S, Singh A. Microearthquake Analysis for reservoir properties at the Prati-32 injection test, The Geysers, California. In: Proceedings, Geothermal Resources Council, Las Vegas, Nevada. 2014

[2] Hutchings L, Jarpe S, Boyle K, Viegas G, Majer E. Inexpensive, Automated Microearthquake Data Collection and Processing System for Rapid, High-Resolution Reservoir Analysis. Geothermal Resources Council, Transactions. In: 2011 Annual Meeting; San Diego, CA. 2011

[3] Hutchings L, Bonner B, Jarpe S, Singh A. Micro-earthquake analysis for reservoir properties at the Prati-32 injection test, The Geysers, California. In: Proceedings, Geothermal Resources Council; Portland: Oregon; 2014;**38**: 519-526

[4] Gassmann F. Translated, Berryman, "Origin of Gassmann's equation". Geophysics. 1951;**64**:1627-1629

[5] Berryman JG. Mixture theories for rock properties. In: Ahrens TJ, editor. Rock Physics & Phase Relations, a Handbook of Physical Constants, AGU Reference Shelf 3. Washington, DC: American Geophysical Union; 1995. pp. 205-228

[6] Berryman JG. Seismic waves in rocks with fluids and fractures. Geophysical Journal International. 2007;**171**:954-974

[7] Mavko G, Mukerji T, Dvorkin J. The Rock Physics Handbook: Tools for Seismic Analysis of Porous Media. Cambridge: University; 2009

[8] Berryman JG, Berge PA, Bonner BP. Estimating rock porosity and fluid saturation using only seismic velocities. Geophysics. 2002;**67**(2):391-404

[9] De Siena L, Del Pezzo E, Bianco F. Seismic attenuation imaging of Campi Flegrei: Evidence of gas reservoirs, hydrothermal basins, and feeding systems. Journal of Geophysical Research. 2010;**115**:B09312. DOI: 10.1029/2009JB006938

[10] Foulger GR, Julian BR, Pitt AM, Hill DP, Malin P, Shalev E. Tomographic crustal structure of Long Valley caldera, California, and evidence for the migration of CO_2 between 1989 and 1997. Journal of Geophysical Research. 2003;**5**:995-1022

[11] Foulger GR, Grant CC, Ross A, Julian BR. Industrially induced changes in earth structure at the geysers geothermal area, California. GRL. 1997; **24**(2):135-137

[12] Grechka V, Mazumdar P, Shapiro SA. Predicting permeability and gas production of hydraulically fractured tight sands from microseismic data. Geophysics. 2010;**75**(1):B1-B10

[13] Hatchell P, Bourne S. Strain-induced time-lapse time shifts are observed for depleting reservoirs. The Leading Edge. 2005;**24**(12):1222-1225

[14] Julian BR, Ross A, Foulger GR, Evans JR. Three-dimensional seismic image of a geothermal reservoir: The geysers, California. Geophysical Research Letters. 1996;**23**:685-688

[15] Gritto R, Daley TM, Myer LR. Joint cross-well and single well seismic studies of CO_2 injection in an oil reservoir. Geophysical Prospecting. 2004;**52**:323-339

[16] Daley TM, Schoenberg MA, Rutqvist J, Nihei KT. Fractured reservoirs: An analysis of coupled elastodynamic and permeability changes

from propressure variationion. Geophysics. 2006;**71**:033-041

[17] Gritto R, Korneev VA, Daley TM, Feighner MA, Majer EL, Peterson JE. Surface-to-tunnel seismic tomography studies at Yucca Mountain, Nevada. Journal of Geophysical Research. 2004; **109**(B3):B03310. DOI: 10.1029/2002JB002036 LBNL-50470

[18] Berge P, Hutchings L, Wagoner J, Kasameyer P. Rock Physics Interpretation of P-Wave Q and Velocity Structure, Geology, Fluids and Fractures at the Southeast Portion of the Geysers Geothermal Reservoir. Geothermal GrittoRes. Council, Transactions, 14. In: 2001 Annual Meeting; San Diego, CA. 2001

[19] Bonner B, Hutchings L, Kasameyer P. A strategy for interpretation of microearthquake tomography results in the Salton Sea geothermal field based upon rock physics interpretation of state 2-14 borehole logs. In: Geothermal Resources Council, Transactions, 14, 2006 Annual Meeting; Reno, NV. LLNL, UCRL-PROC-222141. 2006

[20] Zucca JJ, Hutchings LJ, Kasameyer PW. Seismic velocity and attenuation structure of the geysers geothermal field, California. Geothermics. 1994;**23**: 111-126

[21] Menke W. Geophysical Data Analysis: Discrete Inverse Theory. San Diego: Academic Press, an Imprint of Elsevier; 2012

[22] Toksoz MN, Johnston DH, Timur A. Attenuation of seismic waves in dry and saturated rocks: I. Laboratory measurements. Geophysics. 1979;**44**(4): 111-134

[23] Kachanov M, Kachanov B, Tsukrov I. Handbook of Elasticity Solutions. New York: Kluwer Academic Publishers; 2017

[24] Guilham A, Hutchings L, Dreger D, Johnson L. Moment tensor inversions of small earthquakes in the Geysers Geothermal Fields, California. Journal of Geophysical Research. 2014;**119**: 2121-2137

[25] Johnson L. A source model for induced earthquakes at the geysers geothermal reservoir. Pure and Applied Geophysics. 2014;**171**(8):110-142. DOI: 10.1007/s00024-014-0798-7A

[26] Julian BR, Foulger GR. Earthquake mechanisms from linear-programming inversion of seismic-wave amplitude ratios. Bulletin of the Seismological Society of America. 1996;**86**(4):972-980

[27] Johnson L. Source mechanisms of induced earthquakes at the geysers geothermal reservoir. Pure and Applied Geophysics. 2014;**171**(8):1641-1668

[28] Elkibbi M, Yang M, Rial JA. Crack-induced anisotropy models in The Geysers geothermal field. The Northwest Geysers EGS Demonstration Project; California. 2005

[29] Crampin S, Booth DC. Shear-wave splitting showing hydraulic dilation of pre-existing joints in granite. Scientific Drilling. 1989;**1**:21-26

[30] Wu H, Lees JM. Cartesian parameterization of anisotropic traveltime tomography. Geophysical Journal International. 1999;**137**:64-80

[31] Chapman CH, Pratt RG. Traveltime tomography in anistropic media-I. Theory. Geophysical Journal International. 1992;**109**:1-19

[32] Lou M, Rial J. Characterization of geothermal reservoir crack patterns using shear-wave splitting. Geophysics. 1997;**62**:487-494

[33] Boitnott GN, Bonner BP. Characterization of rock for constraining reservoir scale tomography

at the geysers geothermal field. In: Proceedings, Ninteenth Stanford Workshop on Geothermal Reservoir Engineering. 1994

[34] Tarif P, Bourbie T. Experimental comparison between spectral ratio and rise time techniques for attenuation measurement. Geophysical Prospecting. 1987;**35**:668-680

[35] Ingebritsen SE, Manning CE. Permeability of the continental crust: Dynamic variations inferred from seismicity and metamorphism. Geofluids. 2010;**10**:193-205

[36] O'Connell RJ, Budiansky B. Seismic velocities in dry and saturated cracked solids. Journal of Geophysical Research. 1974;**1974**:5412-5426

[37] Boitnott GN. Laboratory measurements on reservoir rocks from the geysers geothermal field. In: Proceedings, Twentieth Workshop on Geothermal Reservoir Engineering, Stanford University, Stanford, California; January 24-26, 1996. SGP-TR-150. 1995

[38] Winkler K, Nur A. Pore fluids and seismic attenuation in rocks. Geophysical Research Letters. 1979; **6**(1):1-4

[39] Gritto R. Chapter 5.0. Towards the Understanding of Induced Seismicity in Enhanced Geothermal Systems, Final Report. Performance Period: January 29, 2010 through May 31, 2014. Sponsored by Golden Field Office, U.S. Department of Energy, Award No. DE-EE0002756. Principal Investigator: Roland Gritto. 2015

[40] Jeanne P, Rutqvist J, Hutchings L, Singh A, Dobson PF, Walters M, et al. Degradation of the mechanical properties imaged by seismic tomography during an EGS creation at The Geysers (California) and geomechanical modeling. Physics of the Earth and Planetary Interiors. 2015;**240**: 82-94

[41] Evans B, Fredrich JT, Wong T-f. The brittle-ductile transition in rocks: Recent experimental and theoretical progress. In: Duba AG, Durham WB, Handin JW, Wang HF, editors. The Brittle-Ductile Transition in Rocks: The Heard Volume. Washington, DC: American Geophysical Union; 1990. pp. 1-20

[42] Kohlstedt DL, Evans B, Mackwell SJ. Strength of the lithosphere: Constraints imposed by laboratory experiments. Journal of Geophysical Research. 1995; **100**:17587-17602

[43] Darot M, Gueguen Y, Baratin M. Permeability of thermally cracked granite. Geophysical Research Letters. 1992;**19**:869-872

[44] Garcia J, Hartline C, Walters M, Wright M, Rutqvist J, Dobson PF, et al. The northwest geysers EGS demonstration project, California part 1: Characterization and reservoir response to injection. Geothermics. 2016;**63**: 97-110

[45] Tokzoz MN, Johnston DH, Timur A. Attenuation of seismic waves in dry and saturated rocks: I. Laboratory measurements. Geophysics. 1979;**44**: 681-690

[46] Johnston DH, Tokzoz MN. Ultrasonic *P* and *S* wave attenuation in dry and saturated rocks under pressure. Journal of Geophysical Research. 1980; **85**:925-936

[47] DeVilbiss-Munoz JW. Wave Dispersion and Absorption in Partially Saturated Rocks [Ph.D. Dissertation]. Stanford University; 1980

[48] Mavkov GM, Nur A. Wave attenuation in partially saturated rocks. Geophysics. 1979;**44**:161-178. DOI: 10.1190/1.1440958

[49] Lowenstern JB, Janik CJ. The Origins of Reservoir Liquids and Vapors from the Geysers Geothermal Field. California (USA): U.S. Geological Survey; 2002 https://volcano.wr.usgs.gov/observatories/yvo/jlowenstern/geysers/Low_Geys.pdf

Chapter 4

Resistivity and Induced Polarization Application for Urban Waste Disposal Site Studies

Andréa Ustra and Vagner R. Elis

Abstract

Environmental impacts caused by urban and industrial waste disposal are one of the greatest worldwide concerns, since contaminants may migrate to the local environment and contaminate soils and groundwater. Geophysical investigations have been widely used in environmental investigations of waste disposal contaminated sites, not only imaging the affected area but also evaluating the evolution of contamination plume in a timeframe. Geophysical studies in contaminated sites consist of detecting and mapping the area affected by a contamination source and providing information related to the groundwater flow and depth of saturated zone and bedrock. Particularly, electrical methods, such as resistivity and induced polarization, can identify the presence of contaminant and help to map and delineate the contaminated area and provide information related to contaminant mobilization and attenuation. Therefore, these methods are a powerful tool for noninvasive long-term monitoring of waste disposal contaminated sites. In this chapter, we will show case studies conducted in Brazil, over different types of municipal waste disposal sites.

Keywords: waste disposal sites, landfills, contamination plume, resistivity imaging, induced polarization

1. Introduction

One of the greatest worldwide concerns is related to the increase and destination of waste volume produced by world's population. Waste disposal imposes a serious risk of contamination of groundwater by the migration of contaminants from the waste site to the local environment. To protect the environment from this type of contamination, waste disposal sites must follow regulation standards.

Modern landfill facilities are specially engineered for the disposal of solid waste. They operate to ensure protection of the environment from groundwater contamination and landfill gas produced by residue degradation. Municipal solid waste (MSW) landfills are areas properly prepared to receive household waste, as well as other types of nonhazardous wastes. Ideally, MSW landfill projects should ensure that landfills: are built in suitable geological areas away from faults, wetlands, flood plains or other restricted areas; include flexible membranes (i.e., geomembrane) overlaying compacted clay soil lining the bottom and sides; collect leachate (formed when rain water filters through wastes) for treatment and disposal; and operate according with standard practices (compacting and covering waste frequently with soil) [1].

However, inadequate waste disposal activities were a frequent routine in many developing countries until recent years. The direct disposal of residues on the ground, without any selection of waste types and without any protection to avoid soil and groundwater contamination, had been a common practice for decades. Inactivated waste disposal sites can continue to contaminate the groundwater, especially when they are located in hydrological vulnerable areas.

Contamination plumes are formed when leachate reaches the local water table and contaminating the groundwater. **Figure 1** illustrates the formation of a contamination plume. Contaminants are diluted into groundwater and are carried through hydrodynamic dispersion along with the groundwater flow.

Leachate typically presents high concentrations of total dissolved solids (TDS), ammonia, organic carbon, chloride, and iron, among other organic and inorganic contaminants [2]. Even though the exact chemical composition of the leachates produced by MSW disposal sites, they can be associated with high electrical resistivity values. This makes geophysical electrical methods ideal to detect contamination plumes generated by waste disposal sites.

The advantages of employing noninvasive geophysical methods over direct sampling are cost and time consuming of analysis. Geophysical methods can therefore optimize investigation campaigns, maximizing the investigated area and minimizing drilling needs. Another important advantage is that geophysical data are real time measurements of the investigated system. However, due to the inherent ambiguity of geophysical methods, it is often necessary to use direct measurements to validate interpretation.

According with Sharma [3], geophysics can assist waste disposal problem by: locating geological features of interest (i.e., faults and contacts); locating aquifers and hydraulic active features for contamination plume detection; and detecting the waste volume and searching for areas appropriate for waste disposal. Electrical methods have been widely used as a tool for environmental investigations. Its application in contaminated site investigations consists in detecting and mapping the affected area and providing information about groundwater flow and saturated zone and bedrock depth. When detecting the impacted area, electrical imaging

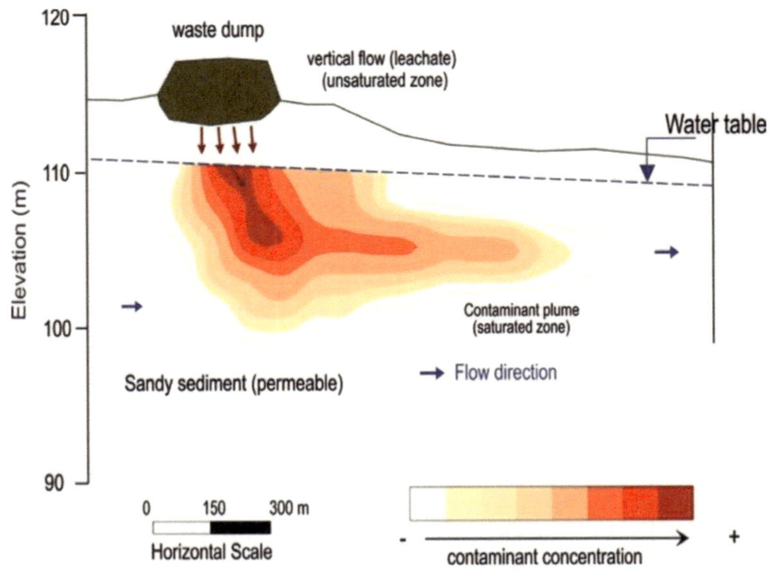

Figure 1.
Contamination plume formed by waste disposal site leachate.

techniques can not only map but also infer contaminant immobilization and long term monitoring.

The direct current (DC) resistivity method is frequently conducted simultaneously to time domain induced polarization (TDIP). In this measurement setup, electrical current is applied as a reversal square wave. When the current injection is "on," the observed voltage, V_c [mV], is measured. When the current injection is "off," the voltage decay curve, $V(t)$, is registered during a period of time $[t_1,t_2]$, from with chargeability, m [dimensionless], can be computed:

$$m = \frac{1}{V_c}\int_{t_1}^{t_2} V(t)dt \tag{1}$$

In the frequency domain, resistivity and induced polarization (IP) methods consist in injecting an alternating current and measuring the amplitude and phase lag between applied current and measured potential, from which in-phase and quadrature components of resistivity, ρ^* [ohm.m], can be calculated: ρ' and ρ''. ρ' represents ohmic conduction (energy loss), whereas ρ'' represents media polarization (energy storage). While m is associated with the intensity of the polarization effect, normalized chargeability, $m_n = m/\rho$ [dimensionless], is considered a direct estimate of polarization, analogous to ρ'' [4].

Distinct mechanisms can generate the polarization response of the media. The IP phenomena are observed when metallic bodies and metallic dispersed particles are present in the subsurface, resulting from differences in ionic mobility in the metallic particles and ions in the pore fluid (electrode polarization). Another source of polarization is ion selective zones formed by clay particles and/or pore throats (membrane polarization). Charge motion along the electrical double layer (EDL) formed at the mineral surface also contributes to polarization (electrochemical polarization) [5, 6].

Resistivity is traditionally applied in waste disposal sites and contamination studies. However, resistivity does not separate different zones in these sites, and low resistivity zones are associated with the whole affected zone, both by wastes and leachate. Johansson et al. [7] and Leroux et al. [8] argue that this limitation can be suppressed by taking into account the normalized chargeability. Despite this seems to be efficient in environments poor in clay content which is not the case of Brazil [9]. According to Slater and Lesmes [4], normalized chargeability is highly influenced by clay content.

Several examples in the resistivity/IP literature report excellent applications of these methods for waste disposal site studies. Bernstone et al. [10] conducted a pre-remediation investigation over a MSW disposal site and identified preferential paths for the leachate. Gazoty et al. [11] obtained a 3D shape of the waste body of the former landfill in Denmark. Maurya et al. [12] investigated the migration of leachate from a landfill in Denmark by 2D and large-scale 3D electrical resistivity tomography. 2D profiles showed variations along the groundwater flow and the plume extension across the flow directions. The 3D model revealed low resistivity variation patterns corresponding to differences in the ionic strength of the landfill leachate.

In this chapter, we will present case studies conducted over municipal waste disposal sites in Brazil. We intend to show the ability of resistivity and IP method in providing useful and fundamental information for investigations of waste disposal and its impact on the environment.

2. MSW disposal site in Ribeirão Preto, SP, Brazil

Ribeirão Preto is a growing population city in Stet of São Paulo, and the water supply is almost completely provided by groundwater water. The quality of the

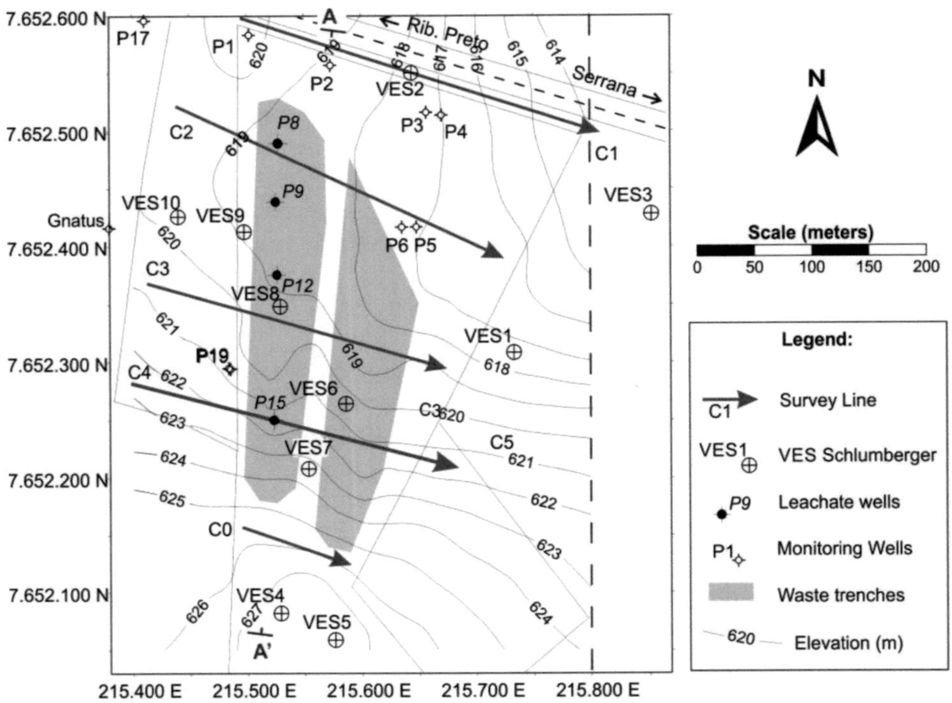

Figure 2.
Local map with survey lines.

groundwater in this region is therefore of critical importance. In this MSW disposal site that operated from 1974 to 1990, wastes were disposed inside two trenches of approximately 15-m deep. **Figure 2** shows the location of the trenches, geophysical lines and soundings, and groundwater wells. The area consists of a surface water divider, and the bedrock is Botucatu Formation sandstone, beneath unconsolidated material composed by sand residual soils and clayey material from Serra Geral basalts [13]. Decomposed basalt is observed at north and south from the trenches, between sandstones and colluvium. In the center, the trench base is in direct contact with the sandstones. The hydrogeological scenario is composed by a deeper aquifer (more than 30 m) within Botucatu sandstone and a shallow aquifer (~ 10 m) sustained by clayey materials originated from basalt alteration. Monitoring wells confirm contamination of groundwater and provide the groundwater flow direction as from southwest to northeast. **Figure 3** presents a geological session of the area, based on the geological wells and geophysical data. The suspended aquifer is assumed to form in the north portion of the trenches, being contaminated by leachate, and the main aquifer that is contaminated since the wastes were directly disposed above the sandstone.

In this site, resistivity and time domain induced polarization profile lines were carried out with dipole-dipole array (10 m of spacing and six investigation levels) using Syscal Pro (Iris Instruments). Metallic electrodes were used for current injection, and nonpolarizing electrodes ($Cu/CuSO_4$) for potential measurement. Current was injected in cycles of 2 s, and the IP measurements were recorded with 160 ms delay after current shut off. Data were inverted with the software RES2Dinv [14] generating 2D models that allowed a detailed analysis of the relationships between the natural materials and the trenches filled of waste.

Figure 4 presents the resistivity and chargeability models of line C4. The resistivity session clearly shows the two trenches filled with wastes and leachate marked by low resistivity values (<15 ohm m). Although the horizontal limits of the trenches

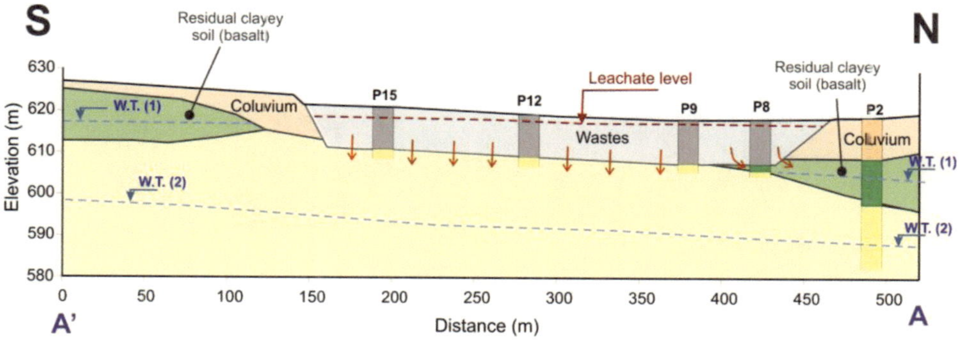

Figure 3.
Profile A–A'.

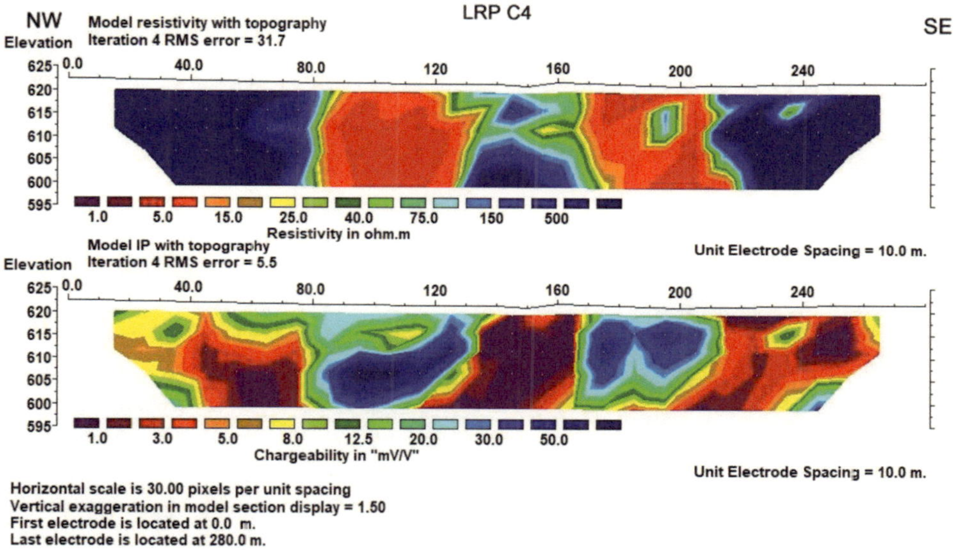

Figure 4.
Resistivity and chargeability sessions of line C4.

are well marked by resistivity, the bottoms of the trenches were not detected. We interpret these results as an indication that the permeable sandstones directly below the wastes as being filled by contaminated water, giving low resistivities. The chargeability session on the other hand, successfully detects the wastes bodies, marked by high chargeability values (>20 mV/V). The sandstone groundwater contaminated by highly saline leachate produces a low chargeability zone (<10 mV/V), identifying the trench base. The same geophysical fingerprint is observed for line C3 (**Figure 5**). The trenches are marked by low resistivity and high chargeability values, while the infiltration zone at the trench bases presents low resistivity and chargeability values. The high chargeability observed in the trenches is explained by the presence of metallic and polarizable material that composes the wastes, whereas the low chargeability signature of the contaminated groundwater is attributed to the decrease of ionic mobility due to increase of solution concentration [12].

Line C1 (**Figure 6**) is located outside and downstream from the trenches. The resistivity session detects the upper portion of the contaminated aquifer, identified by resistivity values lower than 50 ohm m. A polarization anomaly is observed in the position of 135 m along the survey line, produced by the metallic coating of part of the groundwater well P2. The contamination plume is not as well defined by

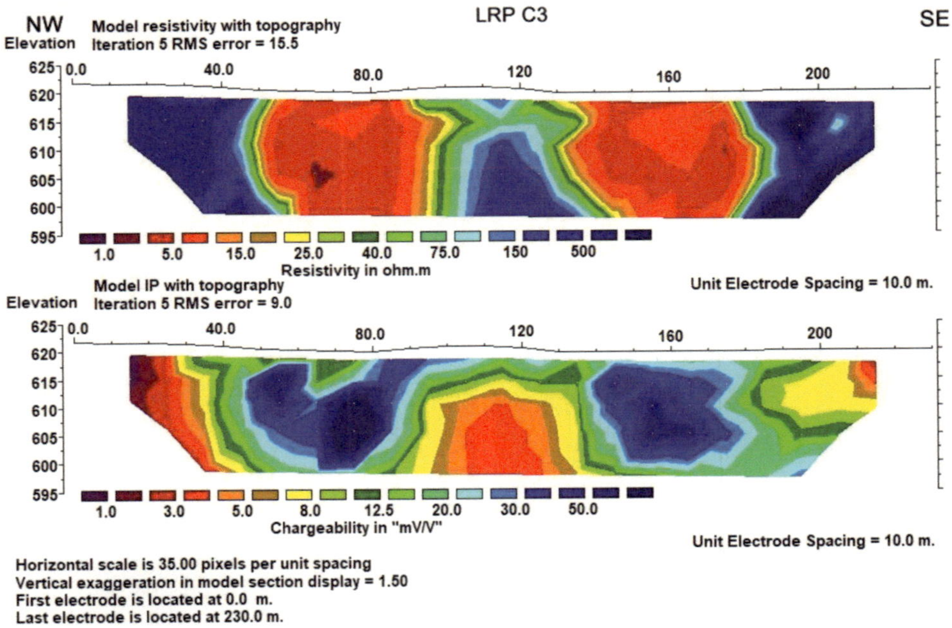

Figure 5.
Resistivity and chargeability sessions of line C3.

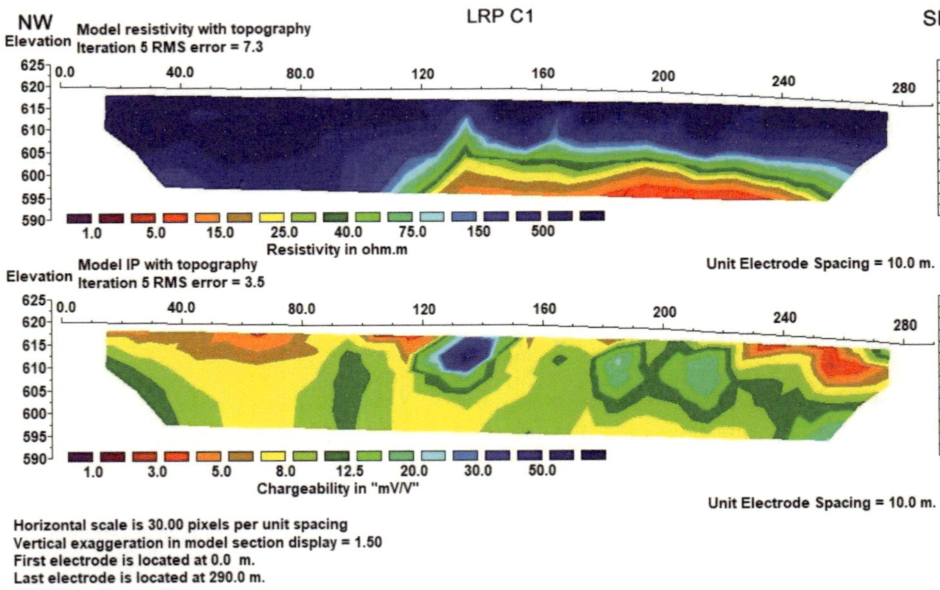

Figure 6.
Resistivity and chargeability sessions of line C1, outside and downstream from the trenches.

chargeability as it is by resistivity, but overall this region shows low chargeability values (8 mV/V). The behavior of chargeability against salinity (and clay content) does follow a linear trend, and its interpretation is not always straightforward. Lithological variations might also be affecting chargeability, competing with the salinity effect.

Table 1 presents chemical analysis of groundwater wells. High TDS values explain the observed low resistivity values inside the trenches. P2 TDS values confirm groundwater contamination. Well P19 (60 m deep) also present high TDS concentration, confirming the contamination of the deeper aquifer.

Monitoring well	Sampled fluid	Electrical conductivity (mS/m)	pH	TDS (mg/l)	COD (mg/l)	Chloride (mg/l)	Sulfate (mg/l)
Maximum allowed value (water)		—	6.5–8.5	500.0	—	250.0	250.0
Gnatus	Water	2.7	6.2	12.0	ND	1.14	6.0
P2	Water	74.0	7.14	1828.0	51.1	—	—
P3	Water	5.4	5.82	26.0	11.0	4.91	3.8
P6	Water	19.5.	6.07	142.0	26.0	8.83	4.8
P17	Water	6.4	6.92	99.0	6.5	4.91	5.2
P19	Water	80.0	10.3	2505.0	130.0	19.0	924.0
P8	Leachate	129.0	6.97	1069.0	91.0	215.0	66.0
P9	Leachate	129.0	6.65	795.0	119.0	262.0	65.0
P12	Leachate	200.0	7.0	817.0	182.0	322.0	103.0
P15	Leachate	1520.0	7.9	9447.0	3640.0	3350.0	145.0

TDS—total dissolved solids, COD—chemical oxygen demand.

Table 1.
Important water parameters of monitoring well analysis.

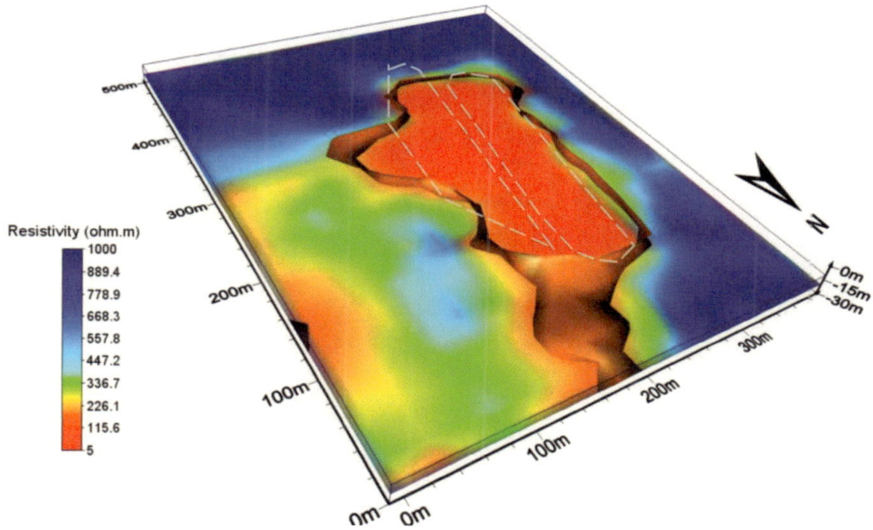

Figure 7.
3D resistivity model of the site, showing the source area (dashed lines) and contaminant plume flowing NE.

Figure 7 shows a 3D resistivity model obtained by the interpolation of all geophysical data. The obtained image shows the contamination source and a contamination zone flowing toward NE to the shallower aquifer.

3. MSW disposal site in São Carlos, SP, Brazil

This study site located in the city of São Carlos, São Paulo State, Brazil is an example of how choosing a disposal are without any consideration for environmental impacts can have disastrous consequences. Wastes were disposed at a natural depression zone produced by intense erosion. There was a small river in this site

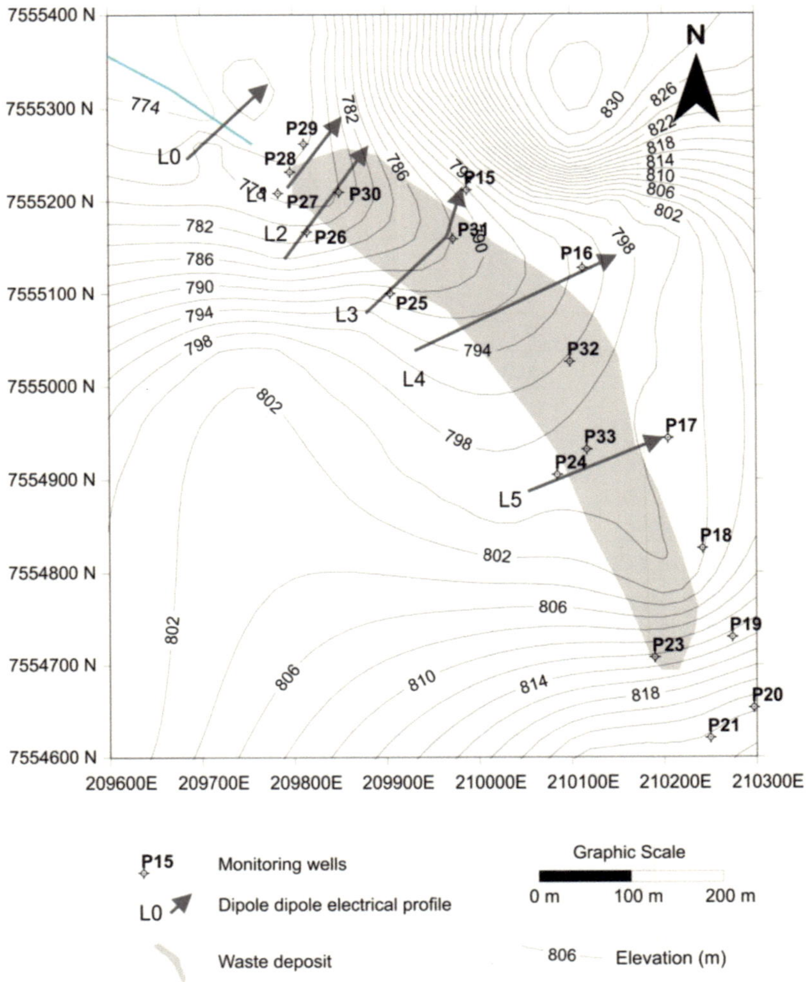

Figure 8.
Local map of waste dump in São Carlos, with geoelectrical survey lines and well positions.

that was covered by the waste dump. The area received industrial, domestic, and hospital wastes over 7 years, and it was then closed with the material being covered by soil in 1996. From the site's topography, the water drain to NW direction is expected, as well as the contamination plume flow in this same direction. Parts of the wastes are deposited in dry soil (above saturated zone), and in the NW part, the trenched bottom is in direct contact with the aquifer.

The site location is at Paraná Basin's east board, where the sandstones of Botucatu Formation occupy most of the study area. Botucatu Formation is the main geological unit of the most important water reservoir in Brazil, the Guarani Aquifer. The sediments from the area are characterized predominantly as fine sand, punctually more clayey or silty.

For mapping this site were performed five resistivity/IP lines across the deposit and one external to it. Data were collected with Syscal Pro (Iris Instruments). Metallic electrodes were used for current injection, and nonpolarizing electrodes ($Cu/CuSO_4$) for potential measurement. Current was injected in cycles of 2 s, and the IP measurements were recorded with 160 ms delay after current shut off. Data were inverted with the software RES2Dinv.

Figure 8 shows the positions of monitoring wells and geoelectrical sessions in the site. The combination of resistivity and time-domain induced polarization (IP)

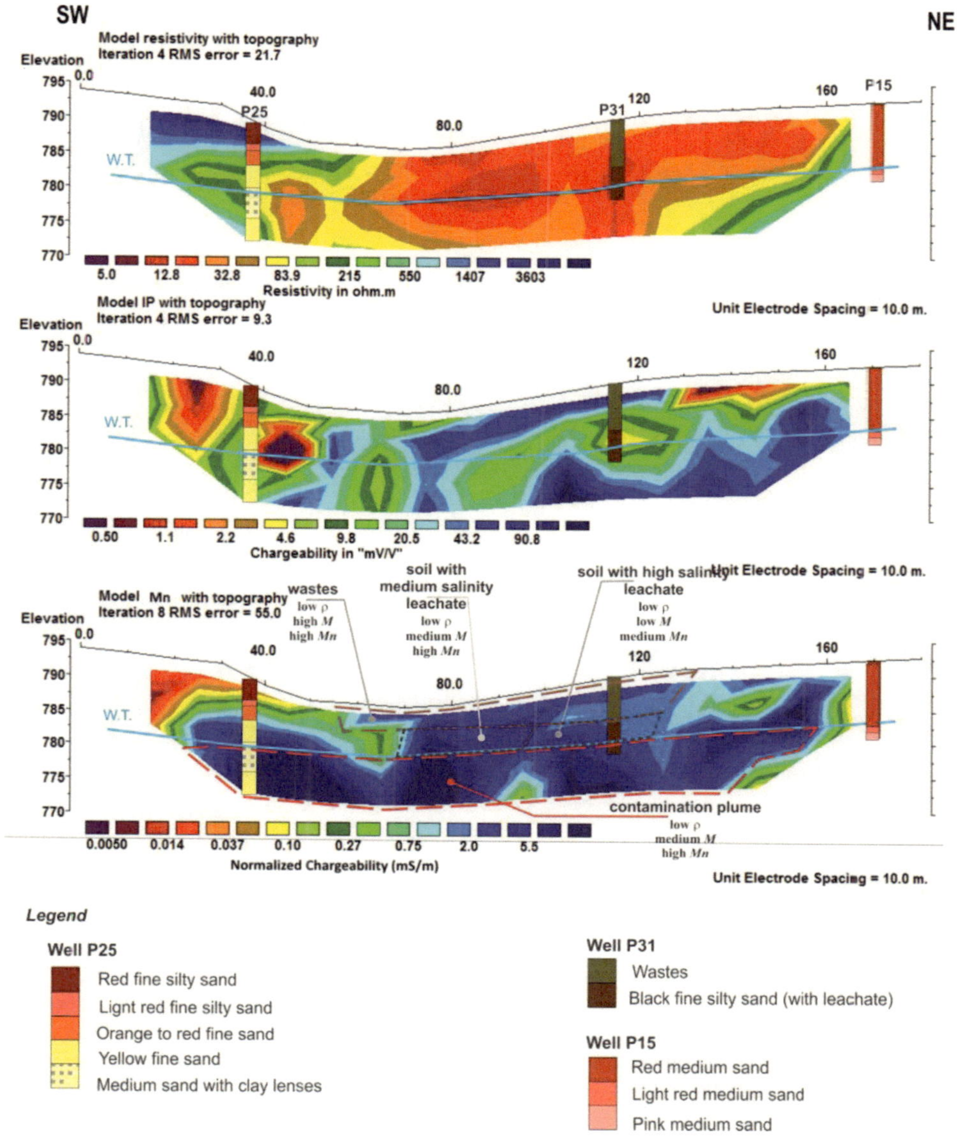

Figure 9.
Resistivity (top), chargeability (middle), and normalized chargeability (bottom) sections of L3 with monitoring wells P25, P31, and P15.

improved the site investigation. We will present results from lines L3 (across the wastes trench), L1 (at the boundary of the trench), and L0 (outside the trench).

Line L3 (**Figure 9**) identifies the area affected by the wastes and leachate, characterized by low resistivity values (<30 ohm m). The waste trench, the contaminated soil, and the contamination plumes are characterized by low resistivity values, and it is not possible to distinguish these different zones. The chargeability session shows the influence of the upper part of the trench (from 60 to 130 m) characterized by high values (>40 mV/V). The leachate attenuates chargeability values, due to its high salinity confirmed by groundwater well Pm31. The contamination plume dilution in the saturated zone produces higher chargeabilities. Normalized chargeability, a direct measurement of polarization, clearly marks the horizontal limits of the trenches and the contamination plume, by high Mn values (from 60 to 130 m, confirmed by visual inspection in the field). Both ρ and Mn show the horizontal spread of the contamination plume.

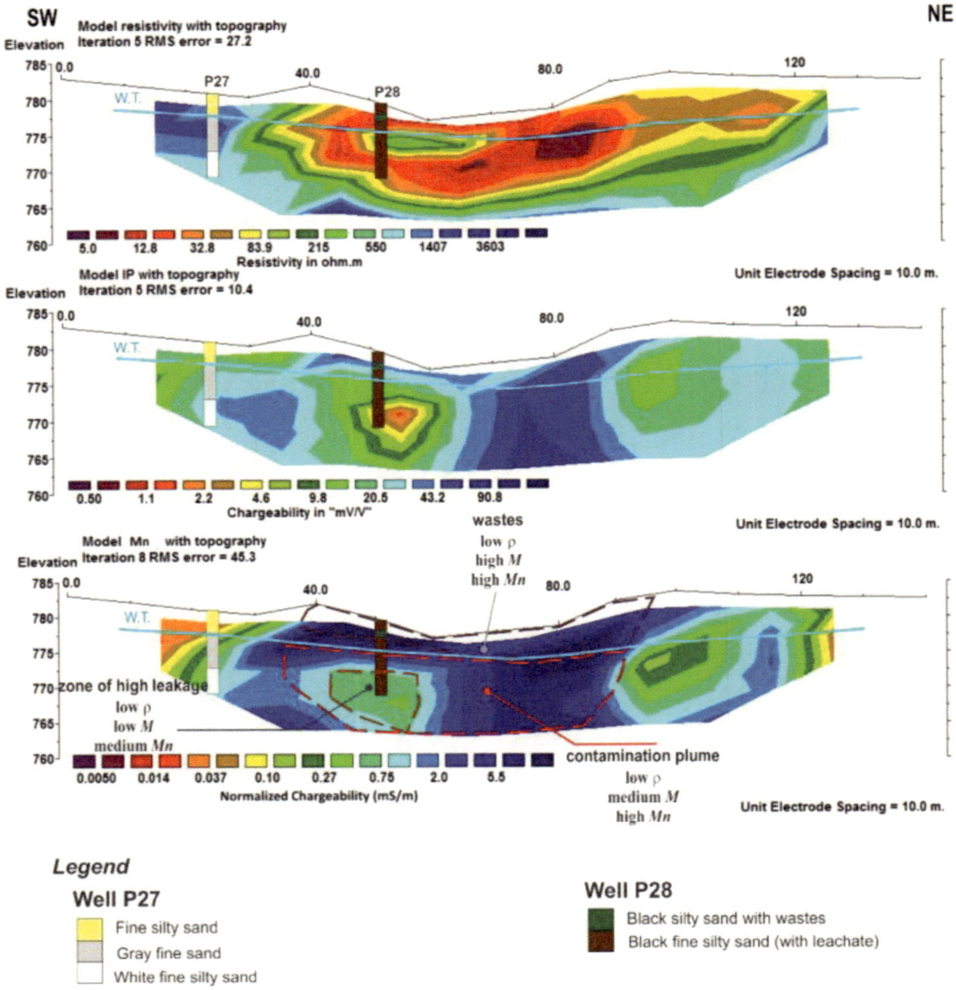

Figure 10.
Resistivity (top), chargeability (middle), and normalized chargeability (bottom) sections of Line 1 and monitoring wells P27 and P28.

Figure 10 presents the results for L1. The resistivity session shows the trench, the contaminated soils, and the contamination plume. Chargeability can limit the wastes between 40 and 90 m and a leachate concentration zone detected by the groundwater well Pm28. Normalized chargeability also identifies the wastes and the contamination plume. In this session, the contamination plume does not spread laterally as L3, which is confirmed by chloride concentrations found in the groundwater well Pm27 (11 mg/l). In the groundwater well Pm25, located in a low ρ and high Mn at L3, the chloride concentration found was 133 mg/l. The line L0 (**Figure 11**) crosses a small river raised from the erosion where the wastes were deposited and the leachate tends to be lixiviated by surface water flow. For this reason, no groundwater wells were installed along this line. The resistivity session shows the water table level very close the surface and low resistivity (50–100 ohm m) in the stream region, characterizing the contamination zone. Normalized chargeability identifies the water table and shows the zones of higher Mn in the stream region. Water sample collected from a small river 50 m always showed a chloride concentration of 64.4 mg/l and an electrical conductivity of 102.8 mS/m. The background values from the stream are to be 0.6 mg/l for chloride concentration and 5.7 mS/m for electrical conductivity.

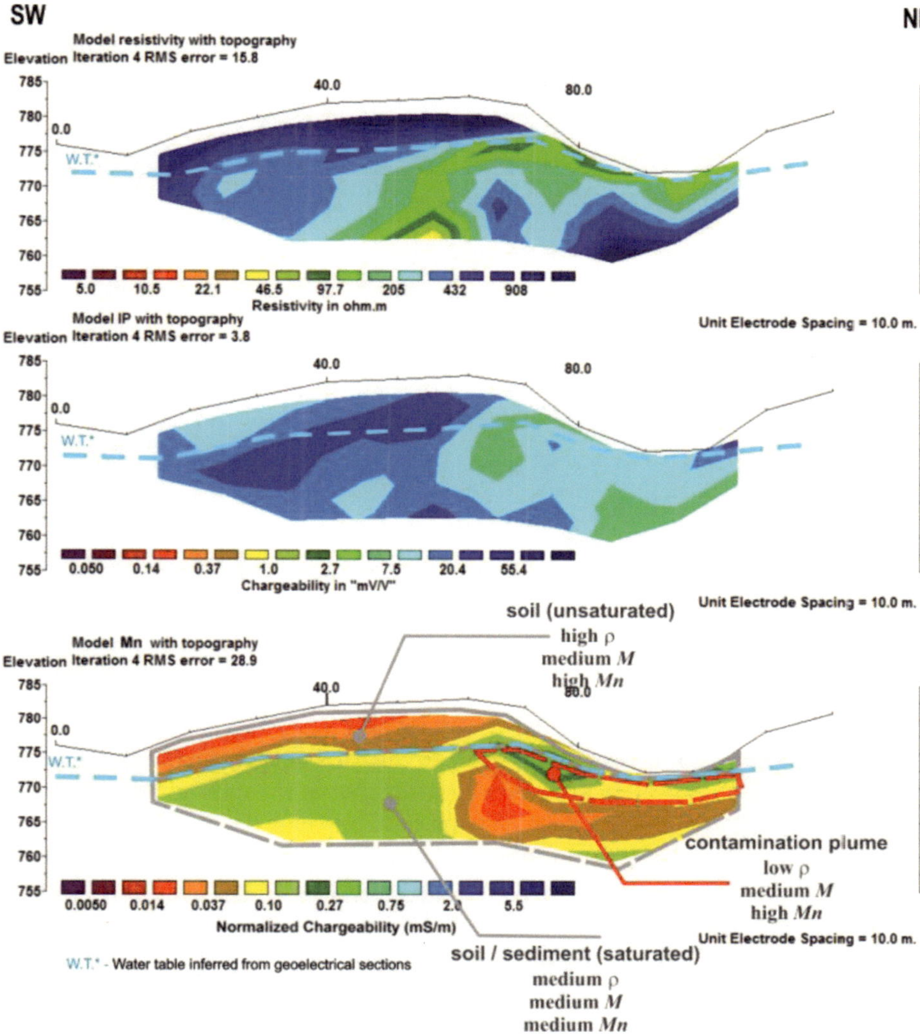

Figure 11.
Resistivity (top), chargeability (middle), and normalized chargeability (bottom) sections of Line 0.

Results from this study show that resistivity successfully identifies the wastes, the contaminated soil, and the contamination plume, being efficient in mapping the affected area. Chargeability is very sensitive to wastes and leachate, but its dependence upon salinity makes its interpretation sometimes complex. Normalized chargeability on the other hand was a more efficient parameter: it increases with fluid conductivity within the fluid conductivity range observed in this study.

4. MSW landfill in ditches in Luiz Antônio, SP, Brazil

In Brazil, sanitary landfills are considered ideal domestic waste disposal sites that are concepted to minimize public health hazard [13]. However, in small towns that produce less than 10 tons of domestic waste per day, waste final destination remains a challenge to the government. In an attempt to improve this situation, São Paulo State Environmental Agency (CETESB) proposed a financial and environmental low cost alternative to waste disposal for towns with less than 25,000 habitants. Landfill in ditches is a technique accepted by the environmental agency

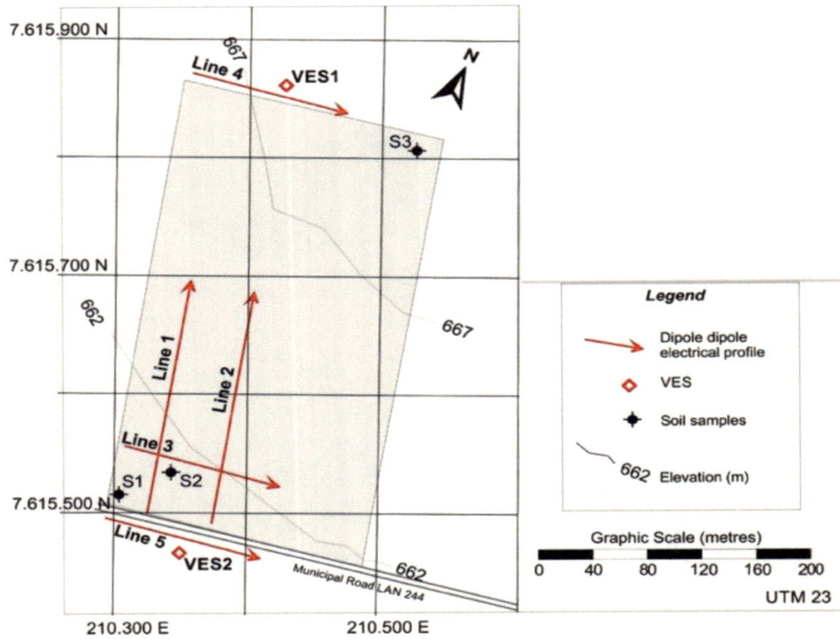

Figure 12.
Local map with survey lines and sample points.

and has been applied for more than 10 years in the São Paulo State, Brazil. The technical concept of landfill in ditches takes into account the natural attenuation capacity of the soil to minimize the environmental impact.

Nowadays, there are a great number of landfills in ditches in São Paulo State, whose simple implementation has been highly contested, making environmental investigation an important task to evaluate this activity impact on the environment. This work investigates the geoelectrical response of a landfill in ditches as an environmental impact evaluation tool. Geoelectrical surveys can help to evaluate the impact on soils and groundwater, leachate migration, and contaminant concentration attenuation. The MSW landfill in ditches in Luís Antônio is active since 1999 and operates in ditches of approximately 70 m long, 8 m wide, and 6 m deep.

Due to project designs, compacted soils are not used as impermeable base for the ditches. Therefore, this type of landfill requires favorable permeability soils and the water table at least 3 m below the ditch base. Luiz Antônio landfill was built on soils of 1×10^{-5} cm/s, and the water table depth is about 12 m.

The geophysical survey was conducted over the waste disposal area (two sessions) and upstream and downstream the ditches (**Figure 12**).

The field survey was conducted with resistivity and time domain induced polarization using dipole-dipole array (5 m of spacing and six investigation levels). The Syscal Pro (Iris Instruments) resistivity meter was used to collect resistivity and chargeability data. Metallic electrodes were used for current injection, and nonpolarizing electrodes ($Cu/CuSO_4$) for potential measurement. IP data were acquired with 2 s integration time with 160 ms delay after current shut off. Data were interpreted with the software RES2Dinv.

Figure 13 shows the line crossing the ditches. Low resistivity (<30 ohm m) and high chargeability (>40 mV/V). No signs of leachate infiltration are observed with the geophysical data. The chargeability value distribution looks complex, but this pattern is a reflex of the waste heterogeneity, and overall, the observed chargeability values are higher than 15 mV/V.

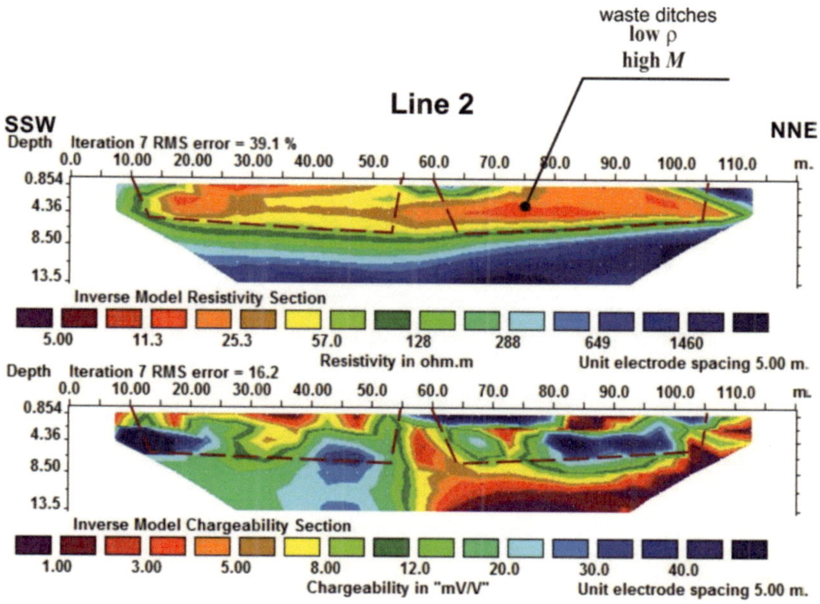

Figure 13.
Resistivity and chargeability sessions of Line 2.

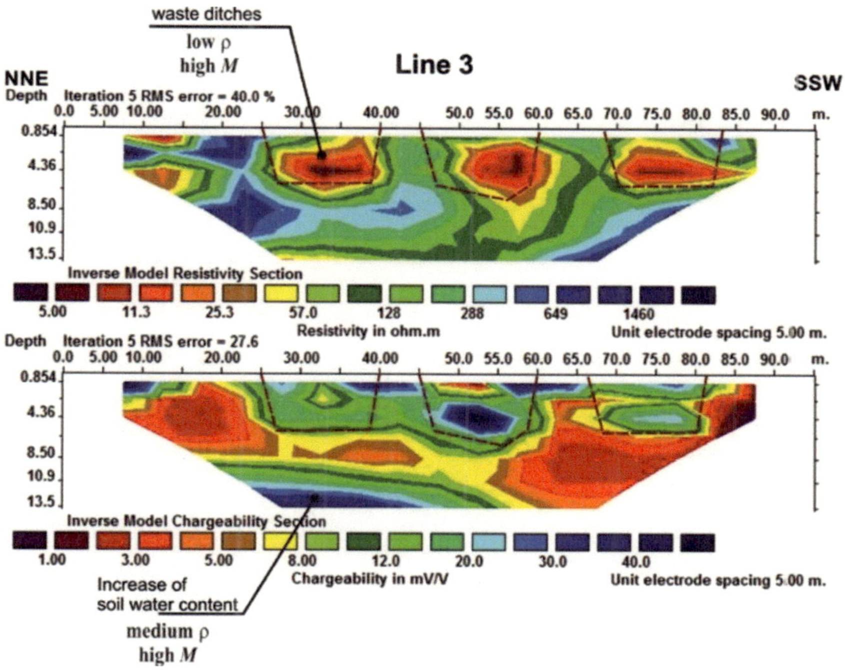

Figure 14.
Resistivity and chargeability sessions perpendicular to the landfill ditches.

Figure 14 shows the resistivity session perpendicular to the three ditches. Each ditch is clearly identified by low resistivity and high chargeability values. A low resistivity continuous feature beneath the ditches is observed, suggesting leachate migration.

To verify the contamination plume, another geophysical line was conducted south, 10 m from the ditches. This line does not show low resistivity features that could suggest a contamination plume. Medium resistivity and chargeability values shown in **Figure 15** are interpreted as higher water content in the porous media.

Soil sampling detected concentration within the legislation limits, even though close to the contamination threshold. Vertical electrical soundings characterized soil horizons, altered and unaltered rock, but the saturated zone was not identified (**Figure 16**).

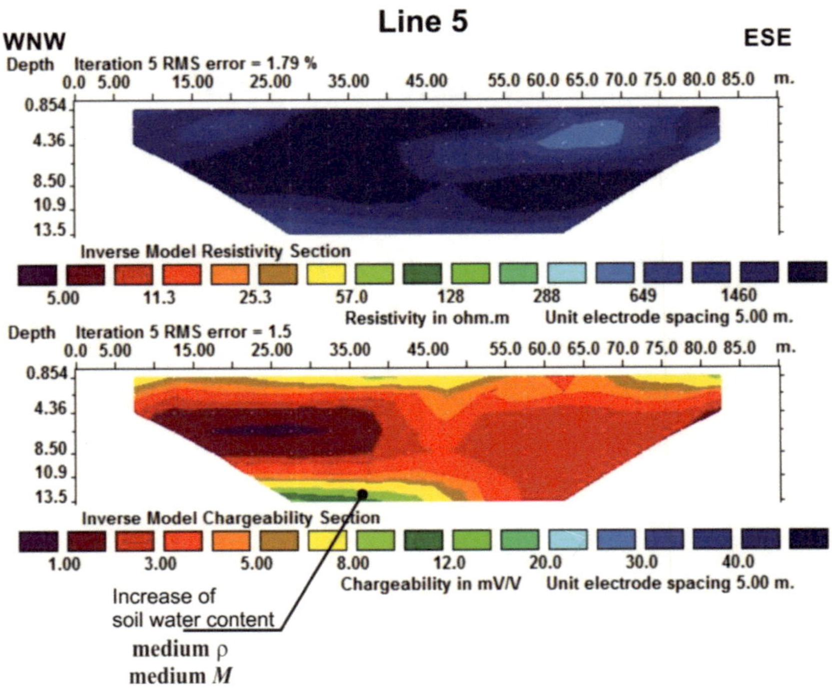

Figure 15.
Resistivity and chargeability sessions south from the trench landfill.

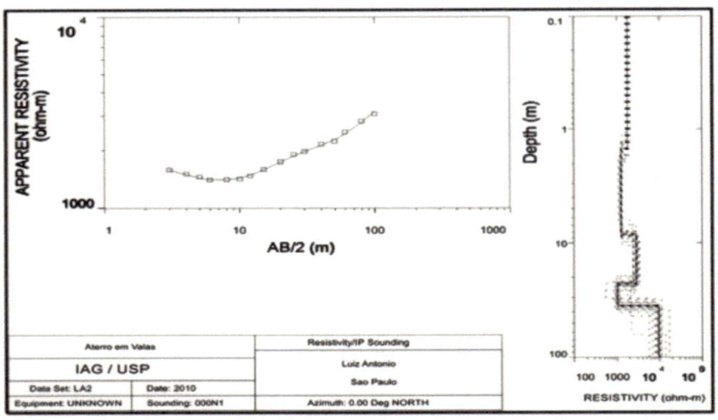

Layer	Depth (m)	ρ (ohm.m)	Interpretation
1	0 – 1.5	1790	top soil
2	1.5 – 8.4	1246	eluviated soil
3	8.4 – 22.4	2836	soil / saprolite
4	22.4 – 34.4	1020	soil / saprolite (increased water content)
5	34.4 –	9550	Basalt

Figure 16.
Interpretation of VES2.

5. MSW sanitary landfill in Bauru, SP, Brazil

The MSW disposal site in Bauru, São Paulo State, Brazil, is the biggest disposal site presented in this chapter, occupying an area of 270,000 m^2. It is a concept like a landfill, and the wastes are disposed in three layers of, approximately, 4 m height. Its bottom has an asphalt emulsion layer above compacted local soil. Digital topographical models show that wastes are disposed over an old valley.

The local geology consists of alluvium, colluvium, alteration soils, and sandstones (Bauru Group). The alluvium is characterized by silt-clayey fine sand and the colluvium by clayey sands. The landfill soil is fine to medium clayey sands. Groundwater flow is from east to northwest, reaching shallow water table depth of 5 m at the base of the landfill [15].

Ustra et al. [16] conducted a geophysical survey over the landfill area and suggested the formation of a contamination plume, based on the resistivity imaging. To investigate the contamination plume, the authors conducted a geophysical survey in the area downstream from the landfill, shown in **Figure 17**.

Chemical analysis data of groundwater sampling from monitoring wells (locations shown in **Figure 17**) showed that the downstream wells (P1, P2, P3, P5A, P5B, P7, P8, and P9) have the highest chloride concentration (12.5–30.5 mg/L, except P9), which is a clear sign of waste disposal leachate contamination. Anomalous sulfate, nitrate, nitrite, and iron (Fe) concentrations in some well downstream groundwater flow are below their maximum permitted value, established by the National Environment Council from Brazil, suggesting that this contamination is in its initial state (low contaminant concentration) and diluted into the aquifer. These parameters, when compared with the natural aquifer values, are considered anomalous even though.

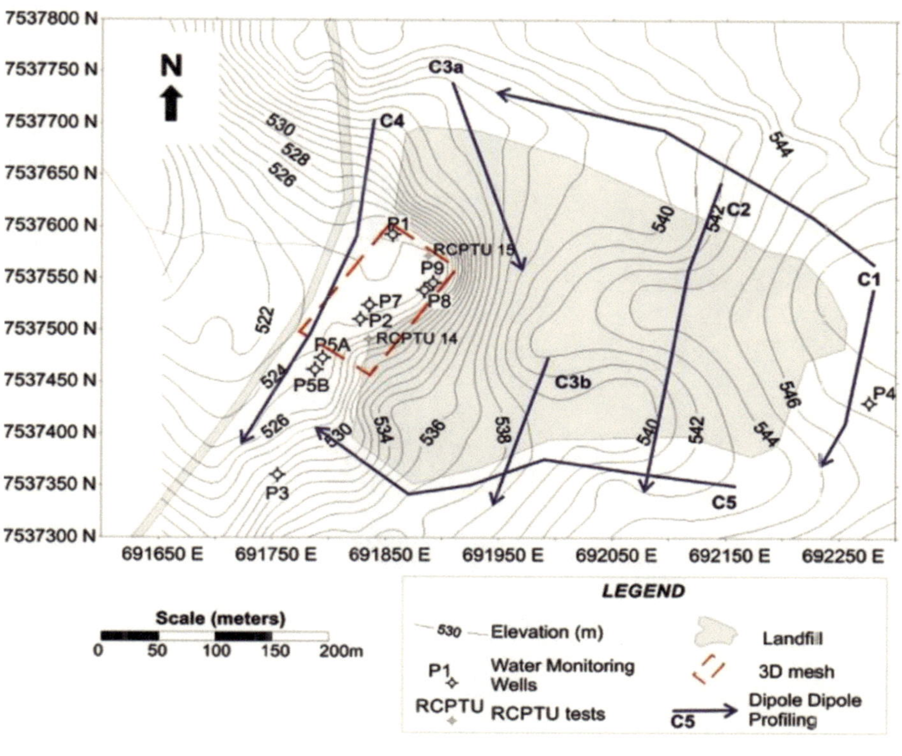

Figure 17.
Topographical map of the landfill with the location of monitoring wells, soundings, and the 3D geophysical survey mesh.

The 3D resistivity model (**Figure 18a**) show resistivity values lower than 100 ohm.m at 10 m deep, known to be n the saturated zone. A conductive feature is observed to start in the landfill (ρ < 20 ohm.m), propagating along the groundwater flow direction. Resistivity values lower than 20 ohm m observed outside landfill are interpreted as contamination plume supported by P2, P7, and P8 chloride anomalous values (mostly in conductive anomaly). However, according to chemical analysis of groundwater contamination, it can only be considered as low contaminant concentration plume. Highest chargeability (~33 mV/V) values are observed in the wastes but also at other regions in the saturated zone (**Figure 18b**). This feature is interpreted as a low salinity contamination plume. Despite low chargeabilities are usually associated with inorganic contamination plumes, the results from this study are in good agreement with Griffiths et al. [17], who observed the increase of chargeability with salinity, over a certain salinity range. The salinity range investigated by Griffiths is exactly the salinity range found in the monitoring well water conductivity. The groundwater flow is best marked by normalized chargeability (m_n > 0.3 mS/m) as shown in **Figure 18c**.

Ustra et al. [16] suggest that the increase of normalized chargeability outside the wastes is a signature of the local contamination plume. This interpretation is in good agreement with Viezzoli and Cull [18] who suggested that in high salinity environments, normalized chargeability is enhanced by clay content. In the case of the landfill, the distribution of the clay content is homogeneous (according with direct sampling), and there is no clay enrichment zone. Ustra et al. argued that in this case, salinity enhancement could highlight the clay content, enhancing the polarization in higher salinity zones that is the contamination plume.

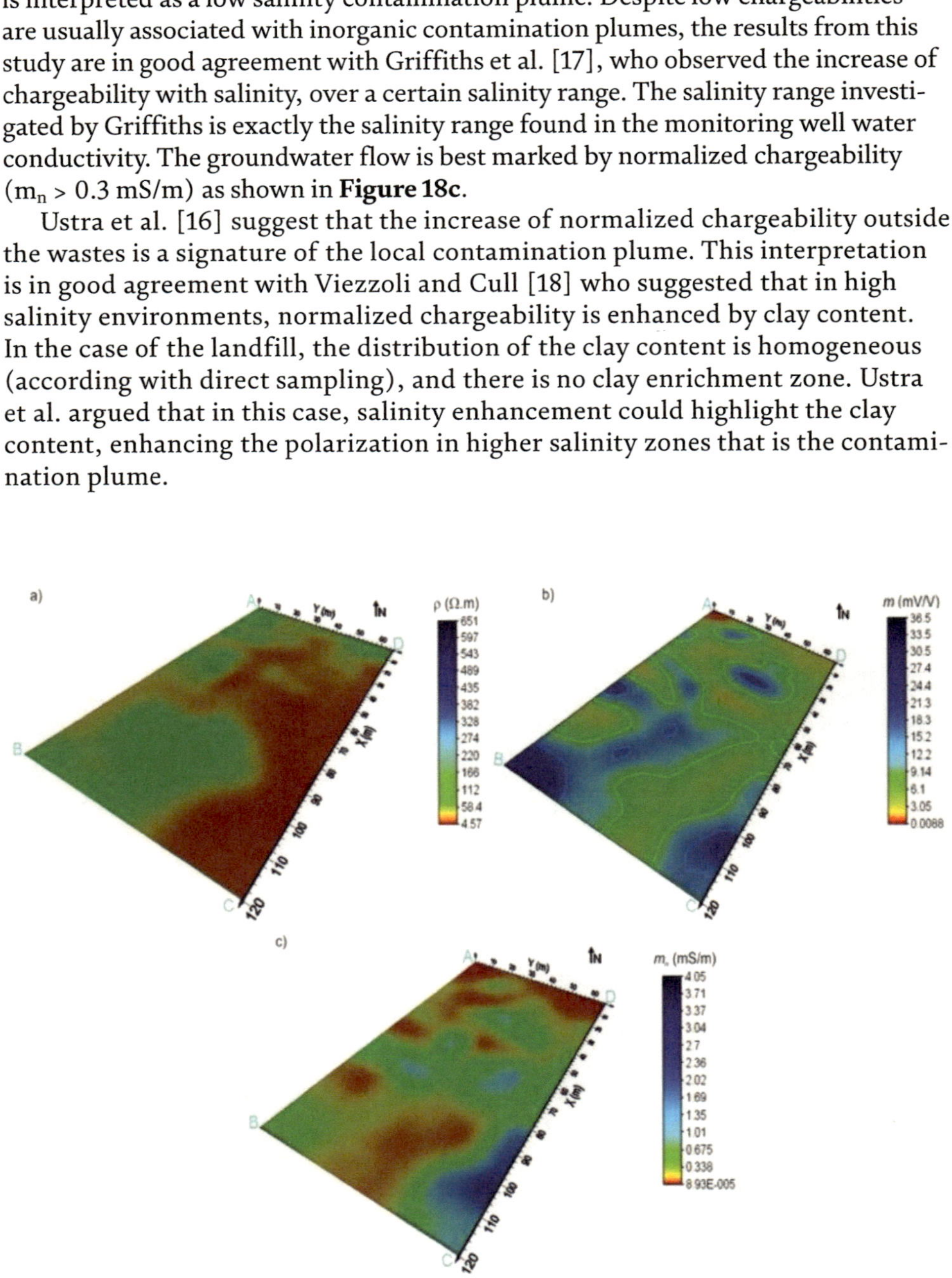

Figure 18.
(a) Resistivity, (b) chargeability, and (c) normalized chargeability models at the depth of 10 m, downstream the landfill.

6. Conclusions

We presented geophysical case studies conducted at MSW disposal sites in Brazil, to show the usefulness of resistivity and IP methods in characterizing these contaminated sites. The examples showed here show different types of waste disposal practice worldwide: direct disposal of wastes on the ground in geologically unfavorable and favorable conditions and proper landfills. We demonstrated that electrical methods are a powerful tool in identifying the presence of wastes in the subsurface and the boundary of the waste volume. Contamination plumes can also be detected with electrical methods, even for low contaminant concentration. Chemical analyses of groundwater are always needed to confirm contamination according to the environmental regulation. When the geophysical signature waste disposal sites are well defined and confirmed, geophysical methods can be used to monitor the temporal evolution of the contamination at the site.

Notes/thanks/other declarations

Thanks to town hall departments from Ribeirão Preto, São Carlos, Luiz Antônio and Bauru, which supported the scientific geophysical investigations at the MWS disposal sites.

Acknowledgements

We thank technicians and graduate and undergraduate student who help the series of surveys presented in this chapter. We also thank Sao Paulo State Funding Agency (FAPESP grant#08/25888-7), National Council for Scientific and Technological Development (CNPq grant #477330/2009-6), and Graduate Support Funding agency CAPES for funding parts of the studies presented here.

Author details

Andréa Ustra* and Vagner R. Elis
University of São Paulo, São Paulo, Brazil

*Address all correspondence to: andrea.ustra@iag.usp.br

References

[1] United Stated Environmental Protection Agency. Available from: https://www.epa.gov/environmental-topics/land-waste-and-cleanup-topics [Accessed: 08-08-2018]

[2] Mirecki JE, Parks WS. Leachate geochemistry at a municipal landfill, Memphis, Tennessee. Ground Water. 1994;**32**(3):390-398

[3] Sharma PV. Environmental and Engineering Geophysics. Cambridge, UK: Cambridge University Press; 1997. 475 p

[4] Slater L, Lesmes D. IP interpretation in environmental investigations. Geophysics. 2002;**67**(1):77-88. DOI: 10.1190/1.1451353

[5] Pelton WH, Ward SH, Hallof PG, Sill WR, Nelson PH. Mineral discrimination and removal of inductive coupling with multifrequency IP. Geophysics. 1978;**43**(3):588-609. DOI: 10.1190/1.1440839

[6] Leroy P, Revil A, Kemna A, Cosenza P, Ghorbani A. Complex conductivity of water saturated packs of glass beads. Journal of Colloid and Interface Science. 2008;**321**:103-117. DOI: 10.1016/j.jcis.2007.12.031

[7] Johansson B, Jones S, Dahlin T, Flyhammar P. Comparisons of 2D and 3D inverted resistivity data as well as of resistivity and IP surveys on a landfill. In: Near Surface 2007—13th European Meeting of Environmental and Engineering Geophysics. Istanbul

[8] Leroux V, Dahlin T, Rosqvist H. Time-domain IP and resistivity sections measured at four landfills with different contents. In: Near Surface 2010—16th European Meeting of Environmental and Engineering Geophysics (Near Surface); 6-8 September 2010. Zurich, Switzerland: Expanded abstracts. P09

[9] Elis VR, Ustra AT, Hidalgo-Gato MC, Pejon OJ, Hiodo FY. Application of induced polarization and resistivity to the environmental investigation of an old waste disposal area. Environmental Earth Sciences. 2016;**75**:1338

[10] Bernstone C, Dahlin T, Ohlsson T, Hogland W. DC-resistivity mapping of internal landfill structures: Two pre-excavation surveys. Environmental Geology. 2000;**39**(3-4):360-371

[11] Gazoty A, Fiandaca G, Pedersen J, Auken E, Christiansen AV. Mapping of landfills using time-domain spectral induced polarization: The Eskelund case study. Near Surface Geophysics. 2012;**10**(6):575-586

[12] Maurya PK, Rønde VK, Fiandaca G, Balbarini N, Auken E, Bjerg PL, et al. Detailed landfill leachate plume aping using 2D and 3D electrical resistivity tomography—With correction to ionic strength measured in screens. Journal of Applied Geophysics. 2017;**138**:1-8

[13] Pejon OJ, Zuquette LV. Importância de estudos geológico-geotécnicos para a disposição de rejeitos urbanos. II Simpósio sobre Barragens de Rejeitos e Disposição de Resíduos. Rio de Janeiro-RJ: CBGB; 1991. pp. 367-377

[14] Loke MH. RES2DINV ver. 3.4—2D resistivity and IP inversion. Penang: M.H. Geotomo Software Users Manual; 2007

[15] Elis VR, Lago AL, Ustra AT, Mondelli G, Giacheti HL. Utilizacão de mapas de resistividade e cargabilidade para posicionamento de sistema de monitoramento geoambiental. In: International Congress of Brazillian Geophysysical Society, CISBGf 2007; Rio de Janeiro, Brazil. 2007

[16] Ustra AT, Elis VR, Mondelli G, Zuquette LV, Giacheti HL. Case study: A 3D resistivity and induced polarization imaging from downstream a waste disposal site in Brazil. Environmental Earth Sciences. 2012;**66**(3):763-772. DOI: 10.1007/s12665-011-1284-5

[17] Griffiths DH, Barker RD, Finch JW. Recent applications of electrical resistivity and induced polarization methods to hydro-geological problems. In: A Survey of British Hydrogeology. London: The Royal Society; 1980. pp. 85-96

[18] Viezzoli A, Cull J. Electrical methods for detection and discrimination of saline groundwater in clay-rich sediments in northern Victoria. Exploration Geophysics. 2005;**36**(3):294-300. DOI: 10.1071/EG05294

Chapter 5

Analysis of Seismic Responses of Rock Massif to Explosive Impacts with Using Nonlinear Methods

Olga Hachay and Oleg Khachay

Abstract

When conducting mining operations in high-stress rock massive, technogenic seismicity is manifested, with forecasting and prevention issues being given much attention in all countries with a developed mining industry. An important role here belongs to the short-term forecast; the methodology for identifying criteria for it is still a problem, both in mining and in seismology. From the point of view of the paradigm of physical mesomechanics, which includes a synergetic approach for changing the state of rock massive of material with different compositions, this problem can be solved with the help of monitoring methods tuned to the study of hierarchical structural media. Changes in the environment, leading to short-term precursors of dynamic phenomena, are explained within the framework of the concept of self-organized criticality, for which the central moments are heterogeneity and nonlinearity. Introduction of the proposed integrated passive and active geophysical monitoring, aimed at studying the transient processes of redistribution of the stress–strain and phase states, can contribute to the prevention of catastrophic dynamic manifestations during the development of deep-lying deposits. Methods of active geophysical monitoring should be tuned to a model of a hierarchical heterogeneous medium and provided with new results of propagation of wave fields in layered block media with inclusions of a hierarchical structure.

Keywords: electromagnetic, seismic, monitoring, impact-threatening rock massive, new analysis, new monitoring methodology, nonlinear, hierarchical model of the environment

1. Introduction

Synergetics is an interdisciplinary science that allows one to imagine the phenomenon of self-organization in the tasks of physics, chemistry, biology, earth sciences, sociology, and other open systems. The term synergetics was proposed in the 1970s by the German physicist Haken [1]. His works had been devoted to the theory of self-organization in various natural systems. At present, important theoretical and experimental results have been obtained, confirming the relevance of synergetic approach for studying the universal properties of open nonequilibrium dynamical systems, cooperative effects in self-organization processes [2]. In recent decades, interest is aimed to study of nonlinear dissipative systems in which a decrease in the number of degrees of freedom that effectively describes them has been observed. Sometimes it is possible to distinguish several degrees of freedom, to which all the others are adjusted. They determine the dynamics of processes and are therefore called order parameters. When studying dissipative systems, their existence allows for a simplified description or construction of an entire hierarchy of simplified models. A decrease in the number of degrees of freedom means that self-organization occurs in the system, i.e., the system has properties that none of the subsystems possesses. To emphasize this circumstance, the theory of self-organization is also called synergetics. The term "dissipative structure" was introduced by I. Prigozin. He and his school helped to establish the connection between the origin of structures, phenomenological models, and representations of nonequilibrium thermodynamics, which played a big role in the theoretical and experimental study of ordering in open systems [3–5]. Removability from equilibrium and nonlinearity can serve ordering in the system. Between order, stability, and dissipation, a nontrivial relationship arises. Ordered configurations that appear outside the stability region of a thermodynamic branch will be called dissipative structures, which can exist only due to a sufficiently large energy or matter flow. The appearance of order in open nonlinear systems is a paradoxical fact.

In equilibrium systems, dissipative processes destroy any order, and thermodynamic equilibrium is established. In nonlinear open systems, together with other processes, dissipative processes affect the type, form, and size of the dissipative structures. It is known that the geological environment is an open dynamic system that undergoes natural and artificial influences at various scale levels, changing its state, resulting in a complex multi-face hierarchical evolution, which is also one of the subjects of the study of geosynergetics [6–10]. The large number of geological systems is open and nonequilibrium, which can exist for a long time only in the mode of pumping energy through them. The termination of the energy flow leads the system to transition to the conservation stage, when the duration of existence is determined by its energy potential due to the accumulated energy at the previous stage. A distinctive feature of open geological systems is their irreversibility and multifunctionality. Using the synergetic approach, it is necessary to clearly distinguish the scale of natural phenomena. Thus, the growth of individual crystals obeys the laws of thermodynamics, and already the morphology of crystals clusters' changes of their forms is determined by the state of the growth medium linked with external influences [11].

Geological environments can be divided into concentrated and dispersed states. Concentrated systems are characterized by their continuity; they represent as a whole on the considered time interval and state by parameters, determining at the first approximation their stationary state. It can be a magmatic chamber, a single fluid system, a block of rocks of a similar composition, a water basin, an oil deposit, or a massive ore body. In concentrated stationary systems, nonequilibrium processes occur aimed for equalizing the thermodynamic parameters characterizing their state. The distributed systems represent a set of autonomous subsystems that are interconnected by heat and mass transfer channels within the framework of irreversible processes, which can be broken down into several stationary states characterized by a constancy of the main control parameters of the process in the chosen time interval. For heterogeneous and complex in composition real geological systems, it is advisable to talk about the equality of not all parameters but only those that determine the macroscopic state of a particular system—its control parameters. In a predominant number of cases, geological systems are non-stationary, since their parameters do not remain unchanged during their existence. The entire development path of such systems is divided into a number of stationary subsystems, characterized by small changes in their parameters at the chosen time interval. Accordingly for each stationary subsystem, a set of stationary processes are fixed in certain structurally real complexes.

At a certain stage of development, an open dynamic system, exchanging matter and energy with the environment, breaks up into a number of subsystems, which in turn can further split into even smaller systems. How to draw boundaries between them if the processes in these systems could take place tens and hundreds of millions of years ago and sometimes billions? The criterion for determining the boundaries of such systems is one of the tasks of synergetics: macroscopic processes in systems where self-organization processes occur in the nonlinear area are carried out cooperatively, in concert and coherently. In the case of geological systems, the boundary will pass along the line of replacement of some structural-material complexes by others, usually mineral aggregates. The basis of the processes of self-organization in open nonequilibrium geological systems is the energy source. If the energy potential does not reach the threshold value, then the processes of self-organization do not occur, but if it is sufficient to compensate for its loss to the external environment, then self-organization processes will appear and spatial–temporal or temporal structures will be formed. The transition of the chaos structure is carried out abruptly. If the energy input into the system is too much, the structuring of the medium stops, and we have a transition to chaos. In any open, dissipative, and nonlinear systems, self-oscillating processes arise, supported by external sources of energy, as a result of which self-organization proceeds [12].

The paradigm of physical mesomechanics introduced by academician Panin V. Y. and his school [13], which includes a synergetic approach, is a constructive tool for studying and changing the state of heterogeneous materials. This result was obtained by this school on samples of various materials. In our studies of the non-stationary geological environment, in the framework of full-scale experiments in real mountain massifs under strong technogenic influence, it was shown that the dynamics of the state can be detected using synergetics in hierarchical environments [14, 15]. An important role in the study of dynamic geological systems is played by a combination of active and passive geophysical monitoring, which can be carried out using electromagnetic and seismic fields. The change in the state of the system on the investigated spatial bases and times is manifested in parameters related to the structural features of the medium of the second and higher rank. Thus, the study of the dynamics of the state and its structure and the phenomenon of self-organization of the array should be led by geophysical methods tuned to the multi-rank hierarchical non-stationary model of the environment [16]. The results obtained from laboratory experiments allowed physicists to propose a model of periodic structural transformations based on a system of nonlinear differential equations that determine the joint evolution of the densities of decaying boundaries, chaotic dislocations, and the boundaries of the emerging structures and also to show that the synergetic scheme allows to describe in a unified way the structurally caused plastic deformation in condensed media [17]. For the fields of plastic deformation and stresses, a system of nonlinear equations is proposed that makes it possible to represent such a regime in agreement with the experimental data. The available results of terrestrial and especially borehole and underground geophysical observations indicate nonlinear manifestations in rock massive during their development.

2. Results of the using theoretical principles for the analysis of active monitoring of mountain massifs under technogenic impact

One of the fundamental problems of mining, which is traditionally attributed to the problems of geomechanics, is the development of theoretical and experimental methods for studying the structure and state of rock massive with a view to predicting and preventing catastrophic phenomena in the mining of deposits. This problem is complicated by the fact that the rock massif is influenced by direct or indirect technogenic impact, which leads to a significant non-stationary of both the structure and the state of it. The ideological inspirer of the search for integrated geophysical and geomechanical approaches to the solution of this problem in the Ural was N. P. Vlokh [18]. An analysis of the manifestations of mountain impacts in the mine workings of the Oktyabrsky deposit of the Norilsk ore site showed that more than 60% of them are confined to tectonic disturbances. Peeling and intensive incineration occur mainly in excavations located outside the zone of influence of clearing works at a distance of 10–12 m from the surface of the tectonic disturbance. In the ores and rocks of medium disturbance, the dynamic phenomena dominate in the areas of interface of the excavations and are accompanied by collapse of the massif. When intersections with excavations of areas of the massif with irregularities have two or more planes of displacement, with a zone of crumpled and fragmented rocks, collapse from the roof and collapses from the sides are observed, accompanied by a dynamic effect and reaching considerable volumes [19]. When conducting mining operations in high-stress rock massive, technogenic seismicity is manifested, with forecasting and prevention issues being given much attention in all countries with a developed mining industry. An important role here belongs to the short-term forecast; the methodology for identifying criteria for it is still a problem, both in mining and in seismology [20]. From the point of view of the paradigm of physical mesomechanics, which includes a synergetic approach to changing the state of rock massive of different material composition, this problem can be solved with the help of monitoring methods tuned to the study of hierarchical structural media [13, 21]. Changes in the environment, leading to short-term precursors of dynamic phenomena, are explained within the framework of the concept of self-organized criticality [17, 20], for which mainly significant moments are heterogeneity and nonlinearity [13].

Within the framework of the IGD SB RAS School, important results have been achieved in studying the state of the rock mass within the framework of nonlinear geomechanics [22] using geophysical methods that have the resolving power to detect the nucleation and decay of self-organizing structures [23].

For the first time, using the electromagnetic method developed in the IGF UB RAS, it was possible to realize the idea of identifying zones of disintegration in an array of rocks within the framework of field studies [24, 25] and to monitor their morphology [26]. The technique used relates to geophysical methods of nondestructive testing. It differs from the previously known methods of semitransparency or tomography by observation systems and the subsequent interpretation method based on the concept of three-stage interpretation [27]. In [28] the first full-scale results on the detection of the self-organization phenomenon in a rock massif with anthropogenic impact and the method of developing stability criteria based on the proposed classification methodology had been represented. These results were obtained on the basis of analysis of several cycles of electromagnetic monitoring of the massif of the shock-dangerous Tashtagol underground mine, conducted in 2000, 2001, 2002, 2003, 2004, and 2005 in a number of excavations located on four horizons at depths from 540 to 750 m in order to reveal the morphology of the zones of disintegration in the near-working space in a rock massif that is under intense man-caused impact and the influence of the natural stress field. In the work [23], studies were conducted aimed at developing criteria for spatial–temporal complex active and passive seismic and electromagnetic monitoring to prevent destructive dynamic phenomena based on 6-year seismic monitoring data conducted by the service of rock shocks in Tashtagol underground mine and the experience gained using the IGF UB RAS systems of induction electromagnetic space–time monitoring on arrays of various material composition before and after mass explosions.

We analyzed the morphology of the structural features of the disintegration zones before a strong dynamic phenomenon during the next cycle of electromagnetic observations at the Tashtagolsky mine during August 2007. On August 9, there was a rock explosion with the energy: lg E = 6.9 in the range of the ort 3 at a level of 16 m below the horizon −280 (**Figure 1**), N = 108 (**Figure 2(a)** and **(b)**). The analysis of the second curve (**Figure 1**) demonstrates the irregularity of the number of weak dynamic phenomena in the array of the entire mine field with energy lgE < 6 in time. So, after a mass explosion on the same day, 42 phenomena were registered, for the next day, already 17, and then this number in the next day is even more reduced. Before the rock impact, there is a significant rarefaction of the number of dynamic phenomena—a zone of calm. During the day, when there was a rock shock after it, 12 weak dynamic phenomena were observed, similar to how it happened on the third day after the mass explosion.

Three days before the rock shock in the ort 4 (**Figure 2(a)** and **(b)**) in the geoelectrical sections of the massif, subvertical discrete structures are found, into which disintegration zones have merged. These structures manifested themselves in a resonant mode at different frequencies and only at one frequency for each of the units. We discovered the same phenomenon earlier during one day at the mine Estyuninsky and SUBR (mine15) [21] (**Figure 3**).

Regarding **Figure 4** we see the comparative data from 2000 to 2007 the distribution of the parameter of the interval intensity Spint (in 2007 the results were given according to the electromagnetic measurements before and after the rock shock) in the bottom of the Ort 2 of the massif, horizon −210 at two frequencies: 5.08 and 20 kHz. According to the classification [29], the state of the array of the Ort 2 was defined as quasi-stable. The obtained results show that, despite the very close location to the site where the rock shock occurred, the massif remains practically in a state described by the gradation quasi-stable. For the period from August 2 to August 13, 2007, the maximum of the parameter *Spint* moved from the fourth interval (3–4 m) to the first (0–1 m) without increasing its amplitude. The emergence of these structures of subvertical morphology is a forecast of a strong dynamic phenomenon; however, in order to determine the place and magnitude of an event, it is necessary to have information on the state of the arrays and their belonging to the appropriate ranks of stability of the array, as was done in [29].

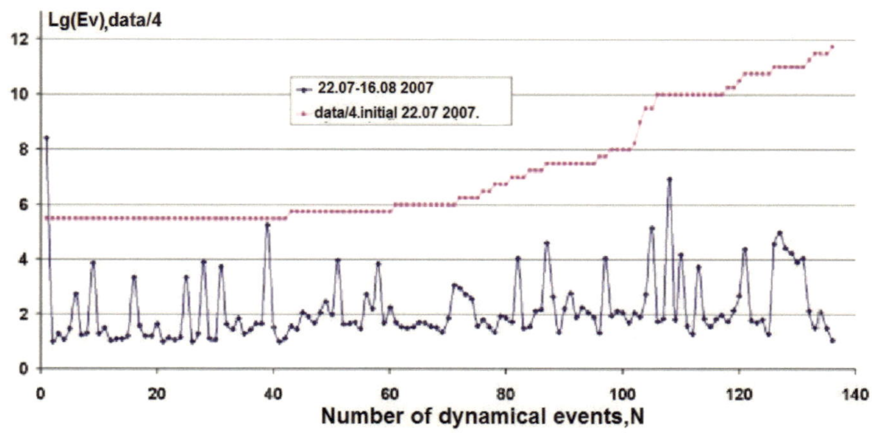

Figure 1.
The distribution of the dynamic phenomena in the Tashtagolsky mine after the mass explosion (N = 1). (Data from the seismological catalog of the Tashtagolsky mine (authors Klimko V. K., Shipeev O. V.)).

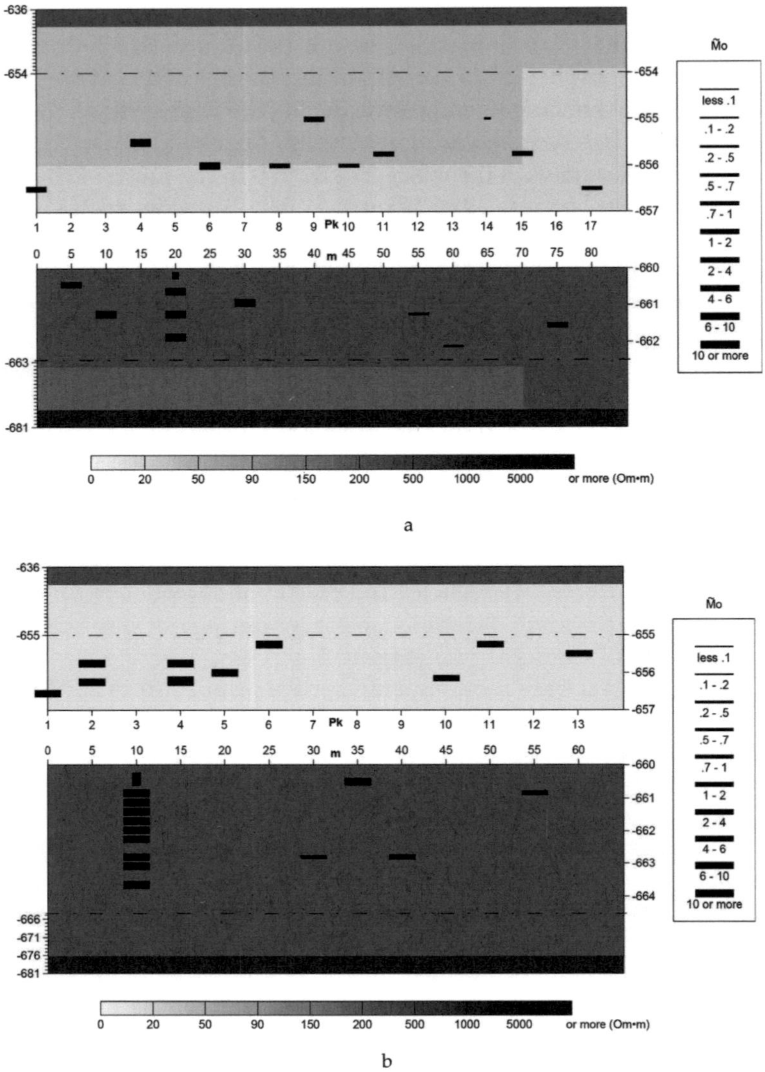

Figure 2.
Geoelectric sections along the profile ort 4, horizon −210, northwestern section. (a) August 6 and (b) August 8, 2007, frequency: 10.15 kHz. Legend: $\tilde{M}_0^i = M_o \times L_o \times 10^3$*, where* M_o *is the coefficient by which the moment of the electric current line is multiplied by the influence of the zone by the parameter of geoelectrical heterogeneity equivalent over the field and which is proportional to the ratio of the conductivity difference of the disintegration zone to the conductivity of the surrounded medium,* L_o *is the length of the current line, and the resistance of the section is given in ohms. Vertical values are given in m (absolute marks), horizontal axes and output length in pickets (pc) and meters.*

Thus, the introduction of the proposed integrated passive and active geophysical monitoring, aimed at studying the transient processes of redistribution of the stress–strain and phase states, can contribute to the prevention of catastrophic dynamic manifestations during the development of deep-located deposits. Methods of active geophysical monitoring should be tuned to a model of a hierarchical heterogeneous medium.

The technology of working out ore deep-lying deposits provides carrying out of preventive and safety control measures. For this purpose, a number of the largest Russian mining enterprises have installed multichannel-automated seismic monitoring systems within the mine fields influenced by underground mining [30].

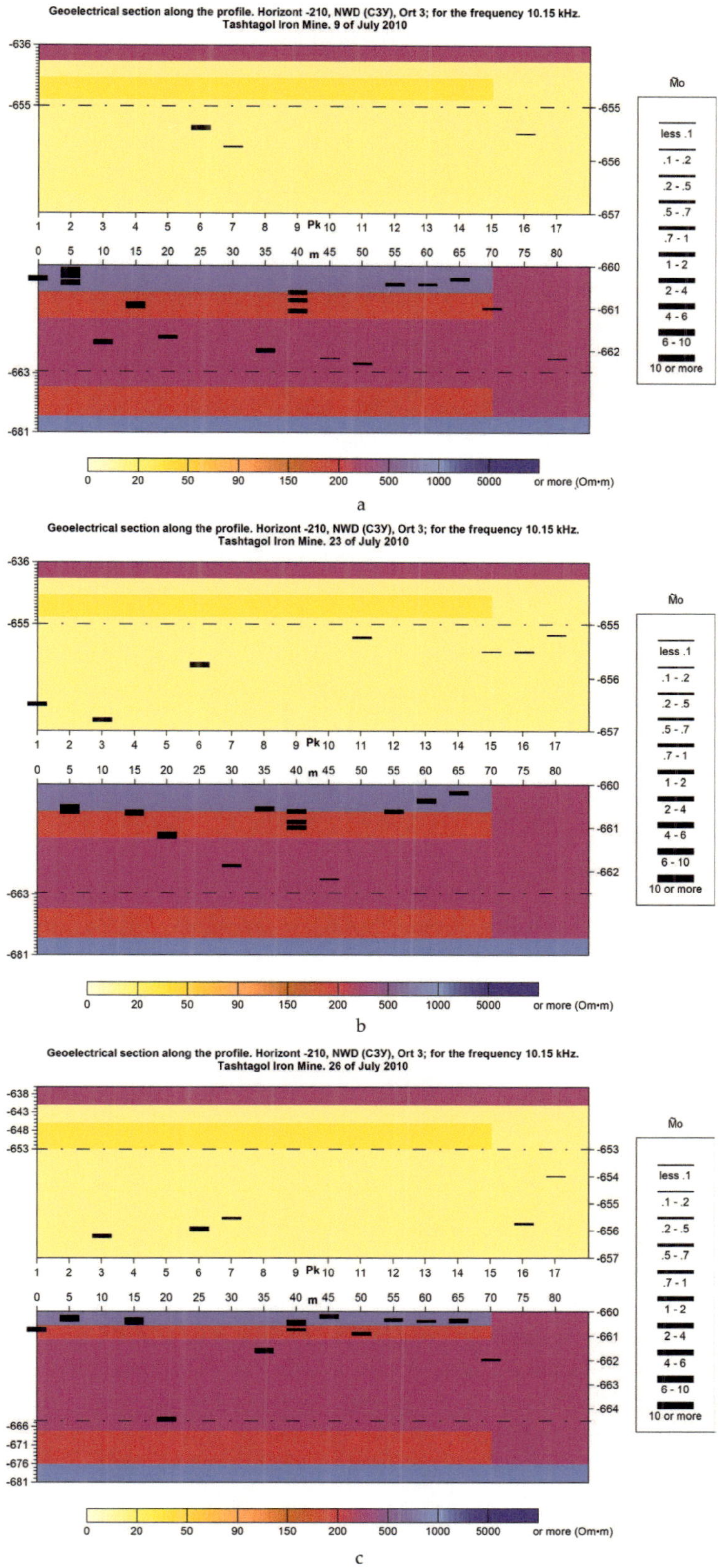

Figure 3.
*(a–c) Geoelectrical section along profile ort 3, horizon −210, northwestern section. (a) July 9, (b) July 23, and (c) July 26, 2010, frequency: 10.15 kHz. The legend is the same as in the **Figure 2(a)** and **(b)**.*

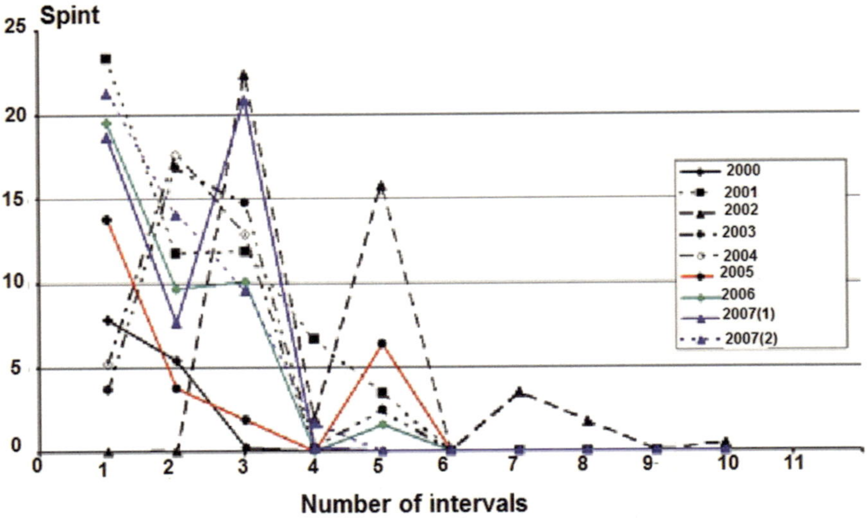

a

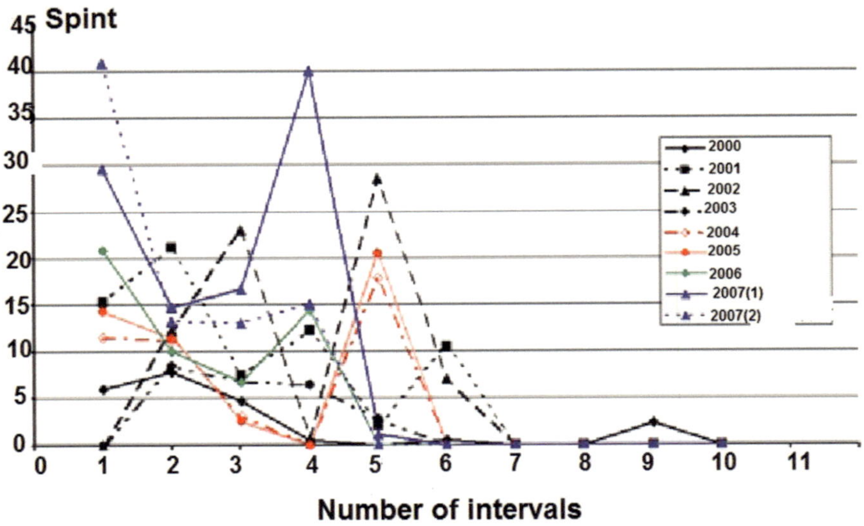

b

Figure 4.
The distribution of the parameter of the interval intensity Spint *for eight observation cycles in the array of Ort 2, horizon −210, Tashtagolsky mine according to active electromagnetic induction monitoring. (a) Observations in 2000–2007 years, frequency 5 kHz. (b) Observations in 2000–2007 years, frequency 20 kHz; 2007 (1) in the legend corresponds to data before rock shock; 2007 (2) in the legend corresponds to data after rock shock. Axes Y-*$\text{Spint}(N,T) = \sum_{i(N)} \tilde{M}_0^i(T)$ *-parameter of the interval intensity of geoelectrical heterogeneities of the second rank, detected by electromagnetic induction monitoring, where N is the number of the interval into which the soil of near-working space is divided and where* $\tilde{M}_0^i$ *is the intensity of the disintegration zone and T is the year of observation [14].*

Analysis of a large database of seismic records of shocks and rock shocks recorded by the Norilsk seismic station at the Norilsk deposit mines using the previously proposed analysis method [31] detects the pulsating seismic energy release from stressed areas of ore and rock massive from the motion of fronts of induced seismicity by the type of oscillating pendulum [32]. In the development of this result, studies were carried out to study the transient process of redistribution of the stressed and phase states of the massif between strong man-made impacts at the

Tashtagol underground mine [33]. We studied the parameters of seismological monitoring as the values of the total energy *lg (Ev)*, extracted by the array of a specific development block in the form of dynamic phenomena after each mass explosion, the values of the absorbed array of the same energy block we define as *lg (Ep)*, and the maximum volume of the mine field where the dynamic phenomena occur from the given mass explosion *lg (Vmax)*.

The transient process of energy release by an array in the form of a response to anthropogenic impact is analyzed—a mass or technological explosion for the realization of a particular technological procedure (cutoff, segmenting, compensation, collapse) in the development block. The analysis of the seismic detailed mine catalog data allows us to draw the following conclusions: when working out a specific block of the array, the entire array of the mine field undergoes a change in the stress–strain and phase states from explosion to explosion; the amount of energy absorbed and delivered by the array is not equal to each other, and therefore energy accumulation takes place in the array; the process of energy release occurs with a delay and depends strongly on the gradient of the energy absorbed from mass explosions; in the array there are zones of dynamic calm; these zones should be monitored using seismic monitoring data using the parameters proposed by us; after it is out of the minimum of the lull, it is necessary to conduct a spatial–temporal active electromagnetic or seismic monitoring within a week or 2 weeks before the technological collapse, in order to identify zones of potential instability of the second rank; these zones may be after a mass explosion, timed to collapse by sources of strong dynamic phenomena; and introduction of the proposed integrated passive and active geophysical monitoring aimed at studying the transient processes of the redistribution of the stress–strain and phase states in the system of testing can help prevent catastrophic dynamic manifestations during the development of deep-seated deposits. These conclusions are based on the analysis of seismological data spatially related to the array of a specific processing unit. However, analysis of seismological data shows that strong dynamic phenomena (rock shocks) can occur in a wider area than the actual block of mining and can be initiated with time lags. Within 9 years from 2000 to 2008, in the mine of the Tashtagolsky mine on four horizons in a number of excavations, active electromagnetic induction monitoring was carried out within the framework of the frequency-geometric technique. On the basis of these detailed data and their subsequent interpretation, a method was developed for estimating and classifying an array of near-working space within the limits of developing its stability in three respects relative to strong technogenic impacts during the development of large and super-large deposits. As a result, a positive verification of the site forecast and evaluation of the magnitude of the destructive dynamic phenomenon in the mine of the Tashtagolsky mine were carried out [34]. As the experience of our studies has shown, the change in the state of the system at the investigated spatial bases and times is manifested in parameters related to the structural features of the medium of the second rank. Thus, the study of the dynamics of the state and its structure and the phenomena of self-organization of the array can be conducted by geophysical methods tuned to the multi-rank hierarchical model of the environment. This conclusion satisfies the principles of the paradigm of physical mesomechanics introduced by academician Panin V. E. and his school [13], which are also a constructive tool for studying the state of the non-stationary geological environment, which is an open dynamic system [14, 27, 35]. The use of an in-plane multi-level induction electromagnetic method with a controlled source and an appropriate processing and interpretation technique made it possible to identify disintegration zones that are an indicator of the stability of the array [26]. The introduction of the integral parameter—the intermittent distribution of the intensity of the disintegration zones—allows us to proceed to a detailed classification of the array in terms of the degree of stability, introduces quantitative criteria for this, and characterizes the stability of the array from the viewpoint of reaching the stationary cyclic position of the maximum of the parameter *Spint* as a function of the distance from the output *Zmax*. Analysis of the variance from the frequency of *Zmax* allows us to introduce additional gradations on the stability of the array in its detailed classification. Comparison with the data of seismological monitoring made it possible to carry out the geodynamic classification of the array using the integral parameter *Sp* [15].

In [35], the possibility of using the mathematical results of the developed physical and mathematical theory of the study of the state of open dynamic conservative and dissipative systems [17, 36] is shown. These include also rock massive during the process of mining. A dynamical system is understood as an object or process for which the concept of a state as a collection of values of certain quantities at a given moment of time is defined and an operator defining the evolution of the initial state in time is specified [36]. If to describe the behavior of a system, it is sufficient to know its state at a finite number of moments of time, and then such a system is called a system with discrete time. As a rule, the control of the state of a rock massif in mines is not continuous but within the framework of observation cycles or at discrete moments of time. To describe its development, differential analogs of differential evolution equations are used. Dynamic systems are divided into conservative and dissipative systems. For the former, the total energy of the system is preserved; for the second, energy losses are possible. In the appendix to our problem, when studying the state of an array that is in the process of working out, the model of the heterogeneous and non-stationary dissipative system is closest. Nevertheless, in the array can be such local parts of it, which will be described by a conservative dynamic model, i.e., model of energy balance. The analysis of the phase portrait of the dynamic system allows us to conclude that the system is characterized during its observation period. So, in conservative systems there are no attracting sets. An attractor is a subset of the phase space P_N, to which trajectories starting in some neighborhood of it incline with time. If a periodic motion exists in a conservative system, then such motions are infinitely large and are determined by the value of the energy under the initial conditions. Attractive sets can exist in dissipative systems. Stationary undamped oscillations for dissipative dynamical systems are not characteristic. However, in nonlinear systems it is possible to have a periodic asymptotically stable motion, in the mathematical image of which is the limit cycle, represented in phase space by a closed line, to which trajectories from some neighborhood of this line are contracted with time. In terms of the shape of the phase portrait, one can judge the characteristic behavior of the system, and the "smooth" deformations of the phase space do not lead to qualitative changes in the dynamics of the system. This property is called the topological equivalence of phase portraits. It allows you to analyze the behavior of various dynamic systems from a single point of view: on its basis, the set of dynamical systems under consideration can be divided into classes within which systems demonstrate qualitatively similar behavior. From the mathematical point of view, the "smooth deformation" of the phase portrait is a one-to-one and mutually continuous transformation of the phase coordinates, as a result of which new singular points cannot appear, and on the other hand, singular points cannot disappear. The earlier results of the study of the phase state of the rock mass [15] indicate that the classification of the massif with respect to its stability and its further control can be very effectively carried out using the parameter *Spint*-interval intensity of second-rank heterogeneities or, according to the terminology adopted in geomechanics, the disintegration zones.

In addition, when using the integrated intensity parameter Sp: ($S_p = \sum_i \tilde{M}_0(x, H)$, where H is the investigated thickness of the massif in the bottom of the hole, the x-coordinate of the center of the zone along the generation, and i is the number of the zone), there is good convergence with seismic monitoring data of the same research area and the active electromagnetic monitoring. Therefore, to construct the phase portrait of the state of the array at various horizons and in the excavations located at various distances from the clearing space, we use as parameters the parameters *Spint* and *d/dt (Spint)*, as well as *Sp* and *d/dt (Sp)* defined for seven cycles of active electromagnetic induction monitoring. By the symbols *d/dt (Spint)* and *d/dt (Sp)*, we mean the difference of consecutive (in time) values; the time interval is 1 year. By a phase trajectory, we mean a discrete set of points in the phase plane defined by phase coordinates in a given time sequence corresponding to the observation cycles. All phase trajectories can be divided into three groups according to the occupied area in the phase plane and the position on the phase plane of the center of gravity of the figure described by this trajectory. By the area occupied by the phase trajectory on the phase plane, we mean the exact lower bound of the set of areas of convex polygons containing a given phase trajectory. The center of gravity of the constructed figure may turn out to be a point of attraction; however, due to the lack of data, this point will be called the center of gravity of the figure described by the phase trajectory. The three groups identified by the new criteria completely coincide with the earlier classification by the parameter *Spint*: stable arrays (mountains −210, ort 4) the smallest area of the figure described by the phase trajectory, quasi-stable (mountains −210, ort 2), (mountains. −350, ort 18) is an intermediate in size area occupied by phase trajectories and unstable (mountain −350, ort 19) is the maximum area occupied by the phase trajectory. Thus, in [37] the thesis that the rock mass is an open dynamic system, the state of which is determined by synergetic properties, was demonstrated quantitatively by analyzing phase portraits using phase coordinates in the form of parameters of the integral and the interval intensity of the heterogeneity zones of the second rank and their difference analogs of time derivatives, determined from the data of active electromagnetic induction spatial–temporal discrete monitoring. To date, the question of the topological equivalence of the constructed phase trajectories, following the definition given above, remains open. Investigating this issue will be possible with an increase in the number of phase data.

3. Using the approaches of the theory of dynamic systems to determine the criteria for changing the regimes of dissipation of real mountain massifs under strong man-made impact

To realize this research, the seismic catalog of the Tashtagol underground mine was used for 2 years from June 2006 to June 2008. As the data, the space–time coordinates of all the dynamic phenomenal responses of the array that occurred during this period inside the mine field, as well as the explosions produced for working out the array, and the value of the energy fixed by the seismic station were used. In our analysis, the entire mine field was divided into two halves: the development of the northwestern section, the areas of the trunks of the western and Novo-Kapital and the outputs from 0 to 14 are designated by us, as the northern section, from 15 to 31 and the southern ventilation and field drifts. The trunk of the southern mine is designated as the southern section. All event responses from horizons −140, −210, −280, and −350 m were taken into account. Explosions were carried out in the southeastern section of mine development at the horizons +70 m,

0 m, −70, on the remaining sections—on the above horizons. As the data, the space–time coordinates of all the dynamic phenomena occurring within the minefield and fixed by the seismic station, as well as their energy characteristics were used.

The phase portraits of the state of the arrays of the northern and southern sections are plotted in the coordinates $Ev(t)$ and $d(Ev(t))/dt$, where t is the time expressed in fractions of the day and Ev is the seismic energy extracted in the array in joules. In this paper, we will first analyze the morphology of the phase trajectories of the seismic response at various successive intervals of time in the southern section of the mine for two reasons: (1) According to the data on the technological and mass explosions produced (**Figure 5**), most of the energy was pumped into the southern part of the mine. (2) At the end of 2007, the one of the strongest rock shocks occurred in the history of the mine happened in the southern sector. **Figure 6 (a)–(h)** shows the evolution of the morphology of the phase trajectories of the array response to technogenic impacts from the middle of 2006 to the middle of 2008. **Figure 6(c)** shows the characteristic morphology of phase trajectories of the response of an array located locally in time in a stable state: there is a local region in the form of a tangle of intertwined trajectories and small outliers from this coil, which do not exceed 10^5 joules in energy. This same feature is manifested in all the figures presented in **Figure 6(d)**, except that at some intervals this ejection exceeds 10^5 joules, reaching 10^6 joules (**Figure 6(d)** and **(e)**) and even 10^9 joules (**Figure 6(g)**). Since the volume of the array under study is the same and we are studying the process of its activation and decay, obviously, there are two mutually dependent processes: the accumulation of energy in the region attracting the phase trajectories and the resonance discharge of the stored energy (e.g., **Figure 6(g)**). It is interesting to note that after this reset, the system returns again to the same region attracting the phase trajectories.

Comparison of the phase portraits of the response of the state of the array before and after the mountain impacts of different intensity and at different time intervals indicate that the volume selected by us in the form of the southern section reacts to the effect exerted on it, similarly, by reflecting a coherent or joint mechanism for releasing the accumulated energy **Figure 8(a-b)**. The first results obtained from the analysis of a detailed seismological catalog from the point of view of the

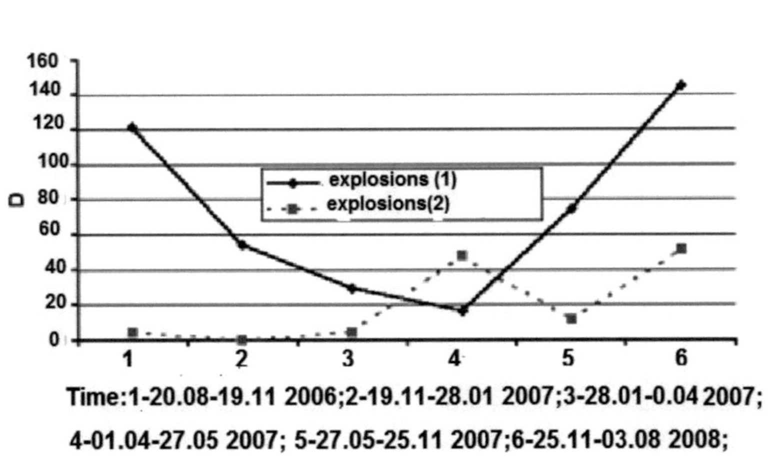

Figure 5.
Distribution in time of the absorbed seismic energy as a result of working out of an array by technological and mass explosions.Axis OY: D = Σlg (Ep (N)), where N is time intervals in days (OX axis) and Ep is absorbed energy from explosions. Explosions (1) provided in the southern part of the mine; explosions (2) provided in the northern part of the mine.

DOI: http://dx.doi.org/10.5772/intechopen.80750

mathematical foundations of synergetics and open dynamical systems possessing the properties of nonlinearity and dissipativity [2, 17, 20, 38] lead us to the necessity of posing a new mathematical modeling problem different from the one previously performed. If in previous productions, the problem of the transition of a system from an ordered state to chaos was investigated; in our case, for our system, the

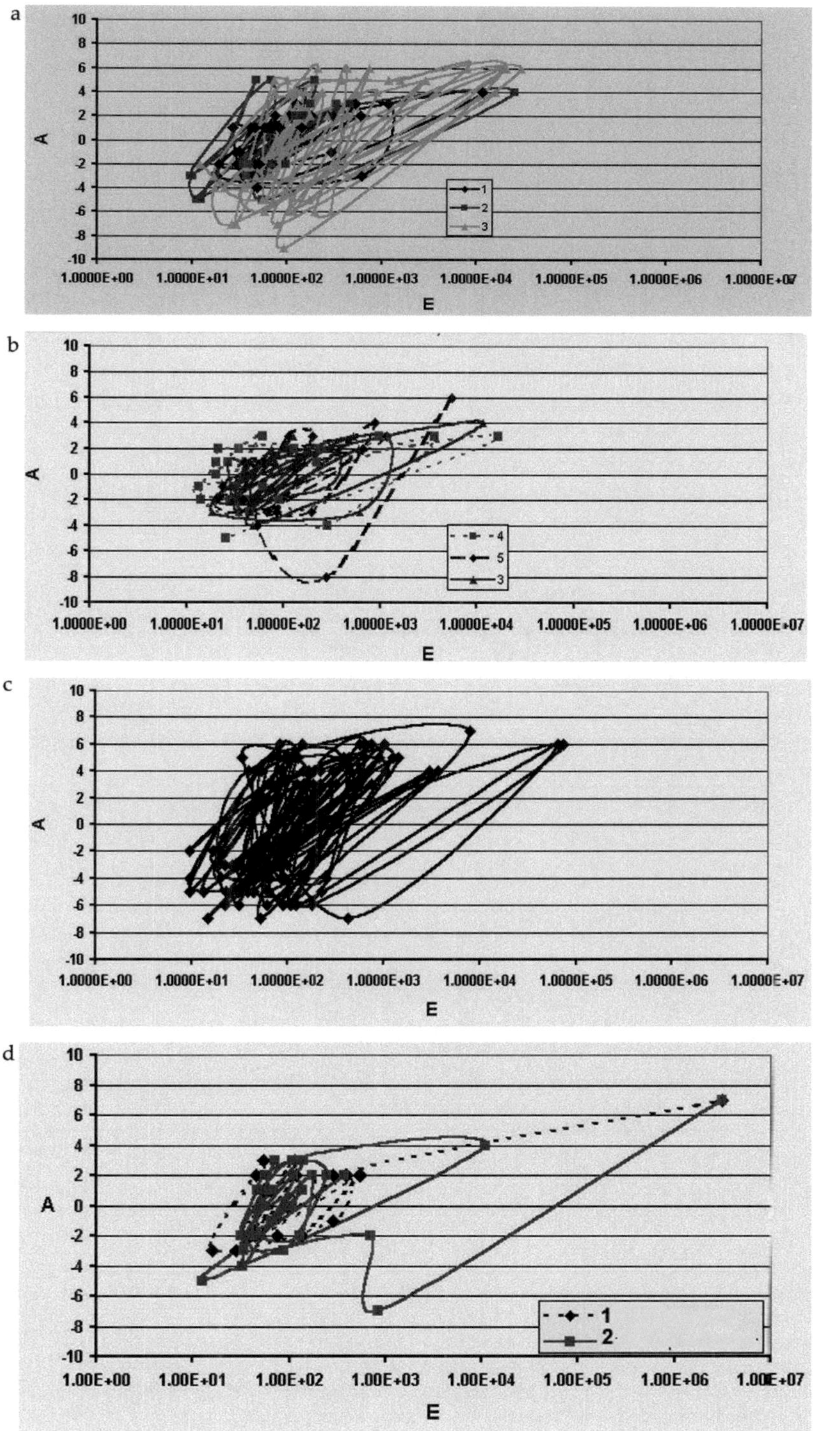

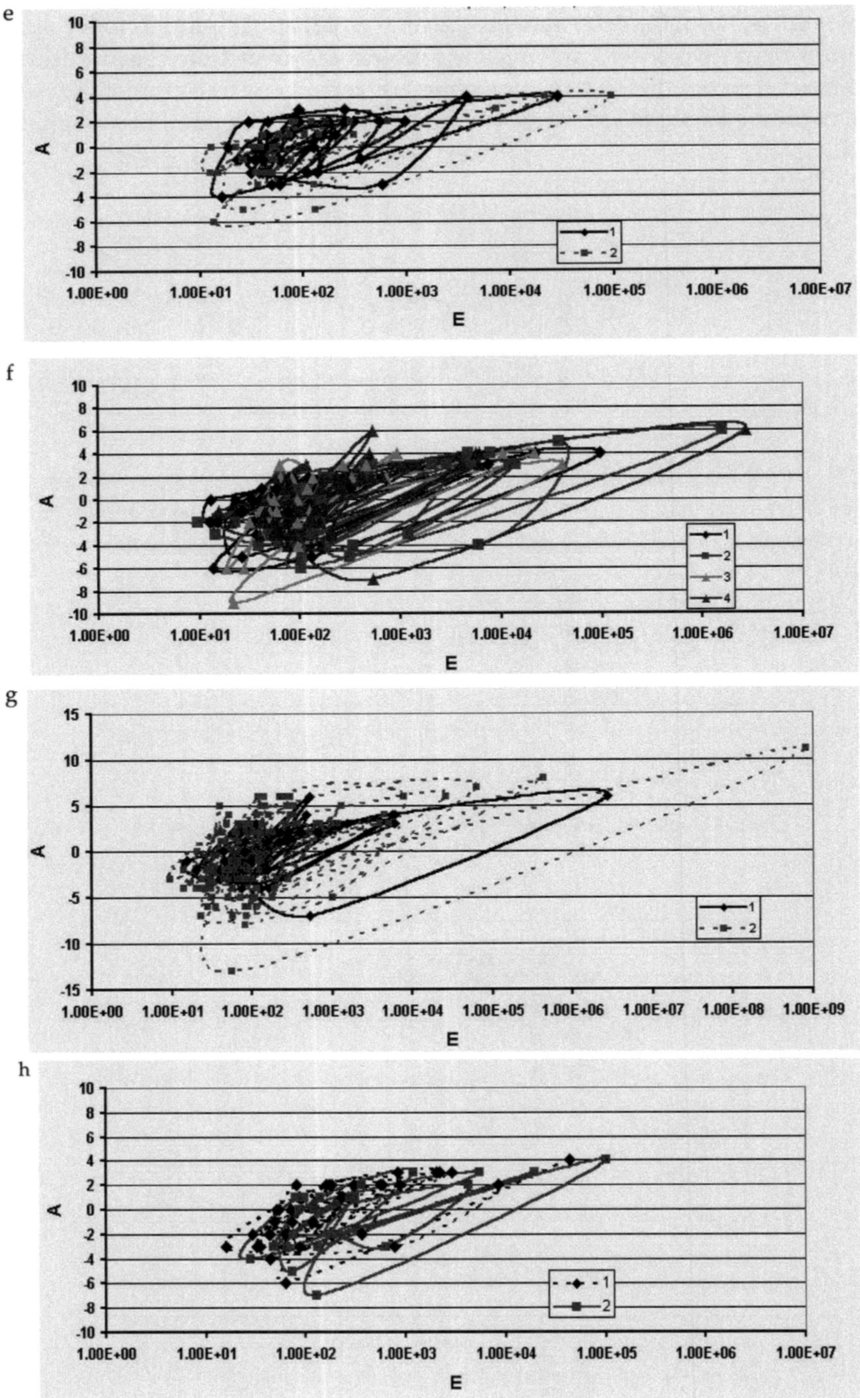

Figure 6

(a–h) Phase portraits of the array state response in consistent time intervals: (a) 5-24.09-29.09, 4-01.10-13.10, 3-14.10-12.11 2006; (b) 1-14.10-12.11, 2-12.11-18.11, 3-19.11-25.11 2006; (c) 24.12-29.12 2006; (d) 1-01.01-28.01 2007; (e) 1-29.01-31.03, 2-02.04-25.05 2007; (f) 1-02.04-25.05, 2-25.06-19.07, 3-22.07-27.09, 4-27.09-24.11 2007; (g) 1-30.09-24.11, 2-25.11-29.12 2007; (h) 1-29.12 2007-21.04 2008, 2-01.06-05.08 2008. The axis OX is the energy allocated by the array in joules at appropriate intervals. The axis OY: A = aLgf, $f = \left|\frac{\partial E}{\partial t}\right|$, a = sign ∂E, *where* t *is time in fractions of a day. It is of interest to analyze in more detail the phase trajectories of the seismic response of the array before and after the strongest impact (**Figure 7(a)–(c)**). The entire process is described by three attractive phase regions: a large number of phase-traversing low-energy region trajectories, which both precede strong energy resonance (**Figure 7(b)**) and follow after a strong energy resonance (**Figure 7(c)**).*

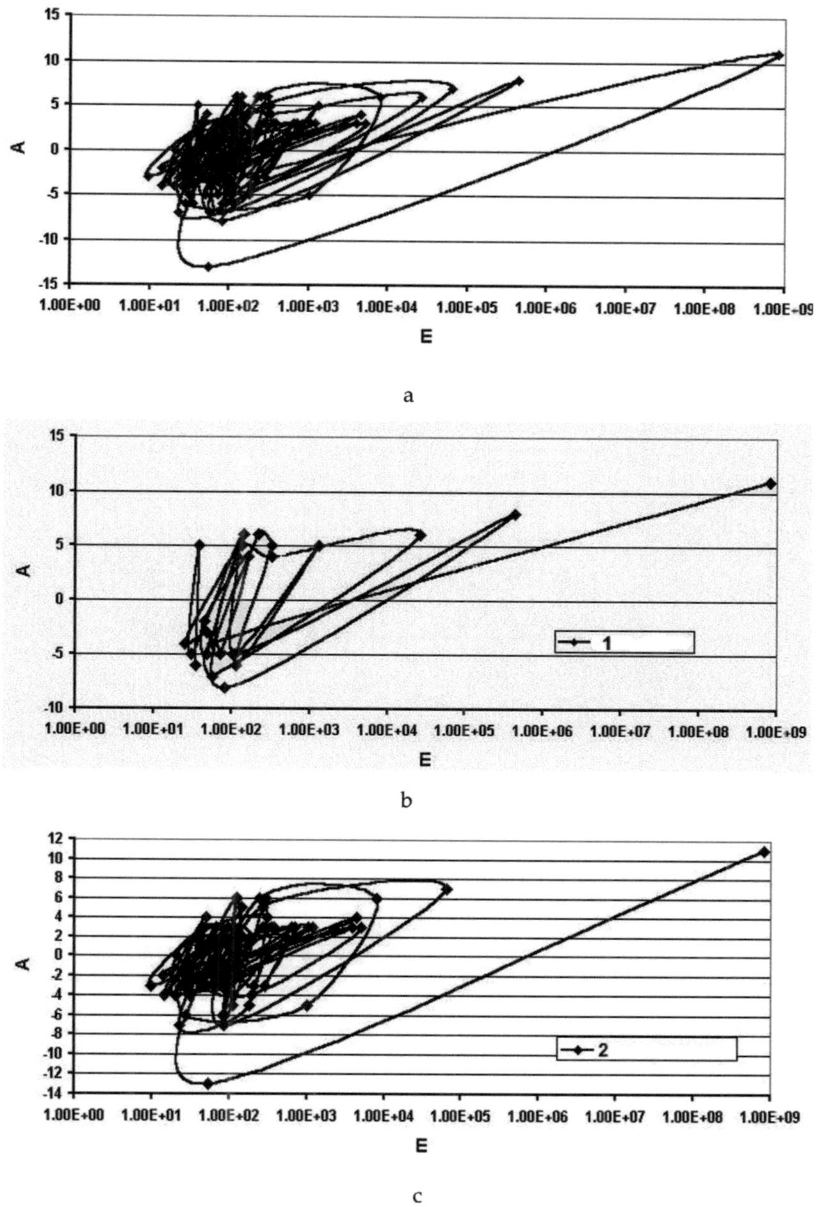

Figure 7.
Phase portrait of the response of the state of the array during one of the most powerful mountain impacts at the Tashtagolsky mine, (a) for a time interval of 25.11-29.12 2007, (b) for a period of time before a rock shock (1), and (c) for a period of time after a rock shock (2). The legend for the axes is the same as in ***Figure 6****.*

chaos of a given level is, on the one hand, a stable state for the system. On the other hand, this parameter is the control for the transition of the system to a state with another parameter, which is catastrophic for it. After the realization of this catastrophe, the system again creates a chaos region with a parameter close in value to the first. This process differs from the bifurcation process, because in the space of the distributions of phase trajectories studied by us, there is an attractive point, in the plane, the extracted energy, and the time derivative of the logarithm of the extracted energy. Thus, further study of the detailed seismic catalog will allow us to formulate the criteria for predicting the behavior of the rock mass from the point of view of the mathematical theory of synergetics [36]. This approach can also be used to analyze seismological data at seismological landfills (**Figure 8**).

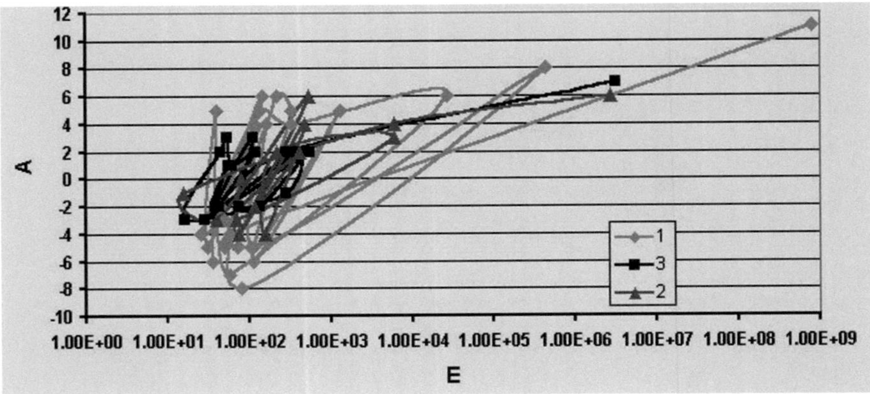

a

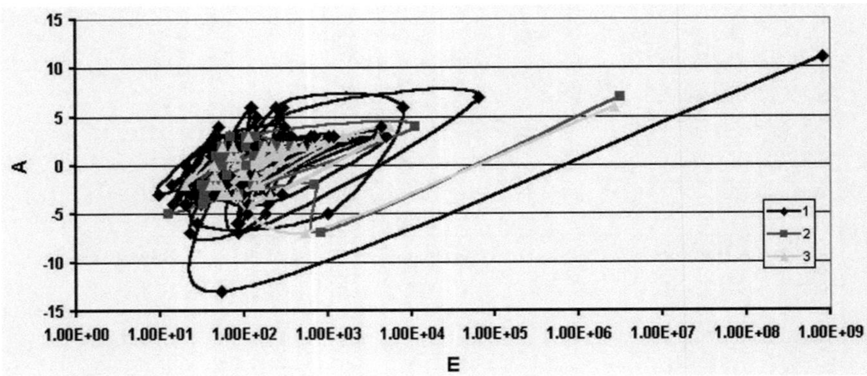

b

Figure 8.
Comparison of the phase portraits of the response of the state of the array before and after the mountain impacts of different intensity and at different time intervals in 2007. (a) 1-25.11, 2-01.01-13.01, 3-27.09-11.10. (b) 1-25.11-29.12, 2-13.01-28.01, 3-11.10-24.11. The legend for the axes is the same as in ***Figures 6*** *and 7.*

4. Conclusion

At present, theoretical results on the modeling of the electromagnetic and seismic fields in a layered medium with inclusions of a hierarchical structure are in demand. Algorithms for modeling in the electromagnetic case for 3D heterogeneity are constructed, in the seismic case for 2D heterogeneities [39–41]. It is shown that with increasing degree of hierarchy of the medium, the degree of spatial nonlinearity in the distribution of the components of the seismic and electromagnetic fields increases, which corresponds to the detailed monitoring experiments conducted in the shock-hazard mines of the Tashtagolsky mine and the SUBR. The constructed theory demonstrated how the process of integrating methods that use the electromagnetic and seismic field to study the response of a medium with a hierarchical structure becomes more complicated. This problem is inextricably linked with the formulation and solution of the inverse problem for the propagation of electromagnetic and seismic fields in such complex media. The problem of constructing an algorithm for solving the inverse problem using the equation of the theoretical inverse problem for the 2D Helmholtz equation is considered in [41], [42]. The explicit equations of the theoretical inverse problem for the cases of scattering of an electromagnetic field (E and H polarization) and scattering of a linearly polarized elastic wave in a layered conducting and elastic medium with a hierarchical conducting or elastic inclusion, which are the basis for determining contours of nonaligned inclusions of the first rank of the hierarchical structure, are written out. Obviously, when solving the inverse problem, monitoring systems that are set up to study the hierarchical structure of the environment should be used as initial monitoring data. On the other hand, the more complex the medium, each wave field introduces its information about its internal structure, so the interpretation of the seismic and electromagnetic field must be kept separately, without mixing these databases.

Author details

Olga Hachay[1*] and Oleg Khachay[2]

1 Institute of Geophysics of Ural Branch of Russian Academy of Sciences, Yekaterinburg, Russia

2 Ural Federal University IEN&M, Yekaterinburg, Russia

*Address all correspondence to: olgakhachay@yandex.ru

References

[1] Haken G. Synergetics. Moscow: Mir; 1980. 404p

[2] Akhromeeva TS, Kurdyumov SP, Malinetsky GG, Samarsky AA. Structures and Chaos in Nonlinear Media. Vol. 2007. Moscow: Fizmatlit; 2007. 488p

[3] Glensdorf P, Prigogine I. Thermodynamic Theory of Structure, Stability, and Fluctuations. Moscow: Mir; 1973. 280p

[4] Nikolis G, Prigogine I. Self-organization in Noneqiulibrium Systems. Moscow: Mir; 1979. 512p

[5] Prigogine I. Introduction to Thermodynamics of Irreversible Processes. Moscow: Science; 1985. 324p

[6] Bulashevitch YP, Khachay YV. Convective stability of the earth's interior with internal sources of heat. Izv.AN. of the USSR. Physics of the Earth. 1975;**12**:13-19

[7] Tikhonov AN, Lubimova EA, Vlasov VK. On the evolution of penetration zones in the thermal history of the Earth. DAN USSR. 1969;**188**(2): 388-342

[8] Trubitsin VP. Tectonics of the floating continents. Bulletin of RAS. 2005;**1**:10-21

[9] Khachay YV. Thermal evolution of the differentiated upper mantle. Geology and Geophysics. 1979;**1**:83-92

[10] Khachay YV. Broadening of the convective cell and the possibility of localization of convection in the upper mantle. Physics of the Earth. 1982;**12**: 19-28

[11] Letnikov FA. Some Problems of Synergetics in the Earth Sciences. Synergetics of Geosystems. Moscow: Moscow Branch of Russian mineralogical Society; 2007. pp. 7-15

[12] Naumov GB. Energy of the Processes of Ore Formation. Synergetics of Geosystems. Vol. 2007. Moscow: Moscow Branch of Russian mineralogical Society; 2007. pp. 15-20

[13] Panin VE. Physical Mesomechanics and Computer Design of Materials. Novosibirsk: Science V.1 SIFR; 1995. 365p

[14] Hachay OA. Geophysical monitoring of the array state of rocks using the paradigm of physical mesomechanics. Physics of the Earth. 2007;**4**:58-64

[15] Hachay OA, Khachay OY. A method for estimating and classifying the stability of a rock massif using the theory of open dynamical systems based on geophysical monitoring data. Mining Information and Analytical Bulletin. 2005;**6**:131-142

[16] Hachay OA, Khachay AY, Khachay OY. Accounting hierarchical heterogeneity of rock massif for prediction of mine seismicity. Open Journal of Geology. Special Issue of Advances in Geomechanics. 2018;**8**(3): 187-200. DOI: 10.4236/ojg.2018.83012

[17] Olemskoj AI, Katsnelson AA. Synergetics of Condensed Matter. Moscow: URSS; 2003. 335 p

[18] Hachay OA, Vlokh NP, Novgorodova EN, Khachay AY, Khudyakov SV. Three-dimensional electromagnetic monitoring of the rock massif state. Physics of the Earth. 2001; **2**:85-92

[19] Egorov PV, Redkin AV. Monitoring of rock impacts during the development of ore deposits with a block structure.

In: Proceedings of International conference “Geodynamics and stress state of the Earth’s interior”. Novosibirsk: Siberian Branch of Russian Academy of Sciences; 2–4 October; 2001. pp. 309-314

[20] Kozyrev AA, Savchenko SN, Panin VI, Maltsev VA. Peculiarities of prognosis and prevention of powerful dynamic phenomena in natural-technical systems. In: Proceedings of the International Conference “Geodynamics and Stress State of the Earth's Interior”. Novosibirsk: Siberian Branch of the Russian Academy of Sciences; 2–4 October 2001. pp. 326-334

[21] Hachay OA, Novgorodova EN, Khachay OY, Kononov AV, Nasedkin VG. Results of geophysical and geological studies at the Estyuninsky mine. In: Proceedings of the International Conference “Deep Structure, Geodynamics, the Thermal Field of the Earth, the Interpretation of Geophysical Fields”. Fourth Scientific Readings in Memory of Y.P. Bulashevitch; 02–06 July 2007. Ekaterinburg; 2007. pp. 197-199

[22] Kurlenja MV, Oparin VN. Modern Problems of Nonlinear Geomechanics. “Geodynamics and Stress State of the Earth’s Interior”. Novosibirsk: Siberian Branch of the Russian Academy of Sciences; 1999. pp. 5-20

[23] Hachay OA. The problem of studying the transient process of redistribution of the stressed and phase states of the array between strong man-made impacts. Mining Information and Analytical Bulletin. 2006;**5**:109-115

[24] Shemjakin EI, Fisenko GL, Kurlenja MV, Oparin VN, et al. The effect of zone disintegrations of rocks around underground excavations. DAN SSSR. 1986;**289**(5):1088-1094

[25] Shemjakin EI, Kurlenja MV, Oparin VN, et al. Discovery № 400. The phenomenon of zonal disintegration of rocks around underground excavations of BI. 1992;1

[26] Hachay OA, Novgorodova EN, Khachay OY. A new technique for detecting zones of disintegration in the near-development space of rock massive of various material content. Mining Information and Analytical Bulletin. 2003;**11**:26-29

[27] Hachay OA, Khachay OY. The study of the self-organization phenomenon in the massive of impact-threatening mines with the use of active electromagnetic monitoring. Mining Information and Analytical Bulletin. 2005a;**5**:92-100

[28] Hachay OA. Phenomenon of self-organization in a rock massif with anthropogenic impact. Physical Mesomechanics. 2004;**7**:292-295. Special Edition, Part 2

[29] Hachay OA. Investigation of the development of instability in an array of rocks using the method of active electromagnetic monitoring. Physics of the Earth. 2007;**4**:65-70

[30] Kurlenja MV, Oparin VN, Eremenko AA. About one method of scanning mine seismological information. Dana. 1993;**333**(6):784-787

[31] Oparin VN, Vostrikov VI, Zhilkina NF, Tapsiev AP, Arshavsky VV. On the pulsating regime of seismic energy release from stressed sections of mine fields. “Problems and prospects for the development of mining sciences”. In: Proceedings of the International Conference. Novosibirsk: Publishing House of IGD SB RAS; 2005. Vol. 1. pp. 7-16

[32] Hachay OA, Khachay OY, Klimko VK, Shipeev OV. The problem of the space-time forecast of the stability of impact-threatening arrays during their development by explosive technologies.

Mining Information and Analytical Bulletin. 2008;**3**:121-129

[33] Oparin VN, Vostrikov VN, Tapsiev AP, et al. On a kinematic criterion for predicting the limiting state of rock massifs from mine seismological data. FTPRPI. 2006;**6**:3-10

[34] Hachay OA. The study of self-organization in hierarchical structures on the basis of spatial- temporal electromagnetic monitoring of impact-threatening rock massive. Russian Geophysical Journal. 2005;**37–38**:24-28

[35] Hachay OA, Khachay OY. Theoretical approaches to the organizing geophysical control systems of the geological environment state that is under technogenic impact. Mining information and analytical bulletin. 2008;**1**:161-169

[36] Malinetsky GG. Mathematical Foundations of Synergetics. Moscow: Publishing House LCI; 2007. 312p

[37] Chulichkov AI. Mathematical Models of Nonlinear Dynamics. Moscow: Fizmatlit; 2003. 294p

[38] Hachay OA, Khachay AY. Modeling the propagation of the seismic field in a layered-block elastic medium with hierarchical plastic inclusions. Mining Information and Analytical Bulletin. 2016;**12**:318-326

[39] Hachay OA, Khachay OY, Khachay AY. New methods of geoinformatic monitoring of wave fields in hierarchical environments. Geoinformatics. 2015;**3**: 45-51

[40] Hachay OA, Khachay OY, Khachay AY. New methods of geoinformatic for the integration of seismic and gravitational fields in hierarchical environments. Geoinformatics. 2016;**3**: 25-29

[41] Hachay OA, Khachay AY. Determination of the surface of a fluid-saturated porous inclusion in a hierarchically layered block environment according to electromagnetic monitoring data. Mining Information and Analytical Bulletin. 2015;**4**:150-154

[42] Hachay OA, Khachay AY. Determination of the surface of anomalously strained inclusion in a hierarchical layered bloc medium according to acoustic monitoring data. Mining Information and Analytical Bulletin. 2016;**4**:354-356

Chapter 6

Application of Seismic Tomography and Geotechnical Modeling for the Solution of Two Complex Instability Cases

Roberto Balia and Pier Paolo Manca

Abstract

The geotechnical instability of sites and artificial structures is a widespread problem that particularly affects heavily anthropized areas and historical cities, and often this instability is linked to human activities and to interventions carried out without adequate preliminary geotechnical studies. The most common procedure for assessing unstable sites includes base studies such as drilling boreholes, shallow excavations, and engineering geology studies. However, more and more often, some geophysical techniques are associated to the above intervention, represent the first phase of assessment, and allow optimizing the possible campaign of excavations and boreholes. Compared to direct surveys, the geophysical ones provide extensive and continuous information, are moderately invasive, and have a remarkably advantageous information-to-cost ratio. In this chapter, we illustrate two examples of characterization of unstable sites. The first case concerns the ancient walls of an Italian city, and the second one deals with the instability of a road slope. In both cases, the geotechnical modeling is also based on the results of preliminary geophysical surveys.

Keywords: slope instability, structure instability, geotechnical modeling, geophysical methods, refraction tomography

1. Introduction

Very often, the instability of sites and artificial structures is attributable to human activities and to interventions carried out without special preliminary geotechnical studies. The most common procedure for assessing unstable sites includes base studies such as drilling boreholes, shallow excavations, and engineering geology studies. More and more often, some geophysical surveys are associated to the above intervention, represent the first phase of assessment, and allow optimizing the possible campaign of excavations and boreholes. The geophysical methods provide extensive and continuous information (usually along sections not necessarily vertical), are moderately invasive, and are convenient from an economic point of view. Two assessment examples of unstable sites are illustrated here. The first case concerns the ancient walls of an Italian city, and the second case concerns a slope on which develops a road.

2. Geotechnical modeling

The combined use of geophysical surveys and geomechanical modeling can make it possible to solve complex geotechnical problems for which geognostic surveys alone prove insufficient [1]. As illustrated graphically in **Figure 1**, the proposed procedure is composed of different phases. These are (1) analysis, (2) measurements, (3) processing, (4) preliminary model, (5) in situ surveys, (6) model revision, and (7) back and sensitivity analysis [2]. The above phases are connected to each other in an iterative way as indicated in the graph, until a satisfactory level of coherence between processing results and direct observations is reached. Actually, there may be cases in which it is difficult to know with precision the complete geometry of the problem, the lithology, and the mechanical properties of natural or artificial materials, as they are not directly investigated. In the present cases, it is possible to carry out iterative evaluations in which the formulation of an initial geomechanical model suggests typology and modality of in situ investigations. In other terms, the initial model integrates and improves the interpretative capacity of the geomechanical model whose processing, in turn, must be compared with direct observations (back analysis or inverse problem). The latter observations could be the observed stability or the instability witnessed, for instance, by easily detectable break surfaces. The last phase of the procedure consists of the sensitivity analysis [3].

Thus, we arrive at the calculation of a safety factor in instability or dimensioning of a stabilization intervention or dimensioning of an additional work that modifies the initial situation without altering the original equilibrium or designing a future monitor system. The two real cases are discussed below.

The first one deals with an excavation in the historic city center for the construction of an underground car parking, in an area flanking a sixteenth century bastion of unknown construction features; the second case is a highway slope affected by rotational instability [4].

In both cases, it is necessary to have a definitive geomechanical model, based on which to design interventions whose effects must be foreseen, starting from initial uncertainties on geometrical and geomechanical aspects. In both cases, the procedure develops starting from direct observations and special in situ investigations. The latter allows to apply a back analysis procedure (or inverse solution) able to highlight critical aspects of the problem and/or to refine the forecast geomechanical model.

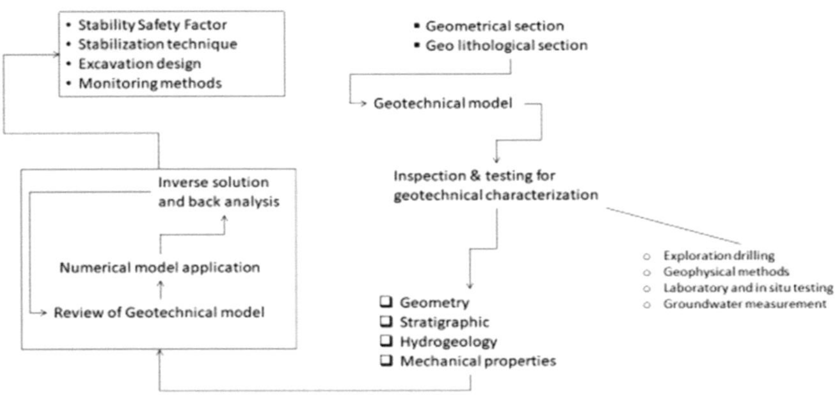

Figure 1.
Flowchart of an iterative analysis modeling—inverse solution.

In the case of the excavation for the construction of the underground car parking, the uncertainties relate to the geometry, composition, and geomechanical properties of the adjacent bastion, obviously apart from the direct observation of the current equilibrium conditions.

In the case of the landslide road, the uncertainties regard the composition of the slope, the stratigraphy and geomechanical properties, and the interaction with a temporary groundwater activated by extraordinary meteoric contributions [5], apart from the direct observation of the landslide in progress.

In both cases, the in situ investigations cannot be exhaustive, due to the discontinuity of the acquired information and to uncertainties implicit in the methodology followed.

Although the nature of the two problems is not the same—feared instability in one case and already occurred instability in the other—they lend themselves to exemplifying the possibility of identifying solutions through the combined action between the geomechanical model and in situ surveys. Among the latter, geophysical-geognostic surveys are those that offer the possibility to acquire the most information possible, by extension and quality.

3. Geophysical investigations

3.1 Geophysical techniques and geotechnical model: the seismic tomography

The most common procedures for recognizing unstable sites include drilling boreholes, shallow excavations, and engineering geology studies. However, more and more often, some geophysical techniques are associated to the above intervention, and usually they represent the first phase of assessment and allow optimizing the campaign of excavations and boreholes. Compared to direct surveys, the geophysical ones provide extensive and continuous information (usually along sections not necessarily vertical), are moderately invasive and have a remarkably advantageous information-to-cost ratio.

Among the various geophysical methods, the seismic ones (e.g., [6]) are the most widely used for geotechnical purposes. The reason is that the velocity of propagation of the elastic (or seismic) waves depends on the density and the elastic properties of the medium in which they propagate. The seismic methods are of different types: first of all the classical seismic refraction, seismic reflection, and the seismic tomography methods. In recent decades, some more methods based on the analysis of the surface wave's dispersion, such as the MASW, multichannel analysis of surface waves; the SASW, spectral analysis of surface waves; and the REMI, refraction microtremor, have been added [7].

Geophysical prospecting methods based on the refraction of seismic waves date back to the 1920s of the last century, mainly in the field of petroleum research. Over time, refraction techniques for oil and gas research have been progressively supplanted by reflection techniques, and their use has shifted to other prospecting fields with objectives falling within the first hundred meters of depth. Actually, the classical seismic refraction shooting based on the analysis and interpretation of the travel time curves has been widely used in the geotechnical field, especially in the study of foundation soils and slope stability. However, before the 1980s of the last century, the processing of seismic refraction data provided approximate models, except in cases of quite regularly layered subsoil. An important step was made with new processing techniques, first the GRM, generalized reciprocal method [8], but still the work was done in terms of seismic rays substantially conceived and represented as broken lines, and only the main refractors could be highlighted. With respect to geotechnical study field, the most important leap, leading to current refraction data processing, has been the operational advent of the seismic tomography [9]. Among the numerous scientific works concerning the topic, the one of White [10] is undoubtedly worth mentioning. As for the inversion algorithm, the most used were FBP, filtered back projection; ART, algebraic reconstruction technique; and SIRT, simultaneous iterations reconstruction technique. In the examples discussed here, the inversion algorithm named ASA—adaptive simulated annealing [11]—has been used. In the seismic tomography, data acquisition is carried out with ordinary energy sources, such as hammers, dropping masses, downhole energy sources, and small charges of dynamite, and with standard receivers such as electromagnetic geophones and piezoelectric hydrophones. **Figures 2–4** show the most common acquisition schemes in seismic tomography. The first one is the classical refraction tomography with both shots and receivers at the ground surface; the second, in **Figure 3**, is the cross-hole tomography where both shots and receivers are placed inside boreholes, better if filled with water; the third, in **Figure 4**, is the up-hole tomography. Referring to **Figure 4**, if the position of shots and detectors is inverted (i.e., the detectors are placed inside the borehole and the shots are placed at the ground surface), the acquisition system is properly named "downhole tomography." It must be underlined that though in the schematic sketch of both cross-hole and up-hole, the ray paths are rectilinear, and this can

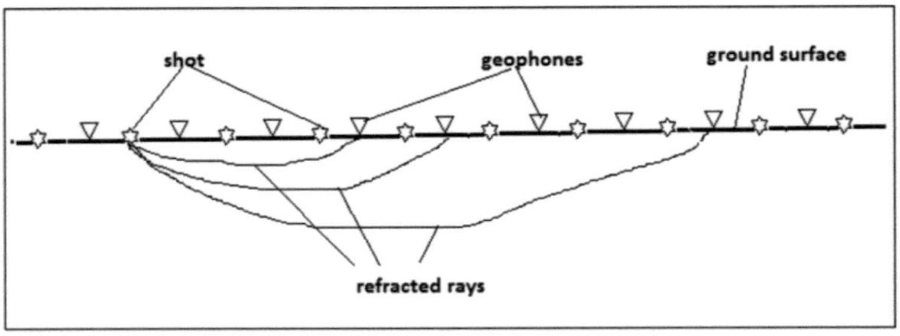

Figure 2.
Acquisition system of the classical seismic refraction tomography.

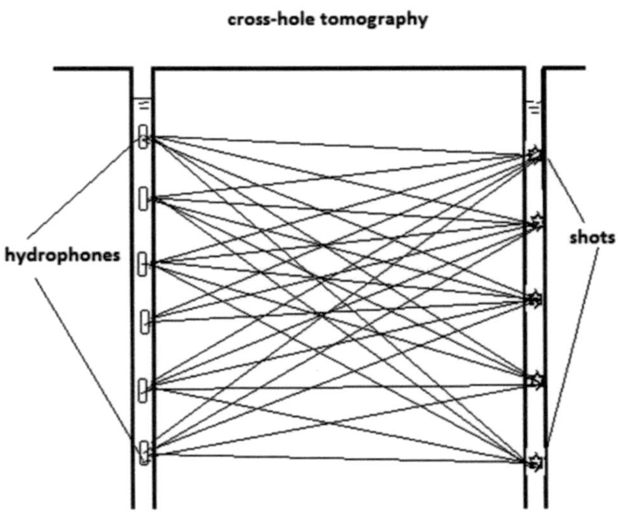

Figure 3.
Acquisition system of the seismic up-hole tomography.

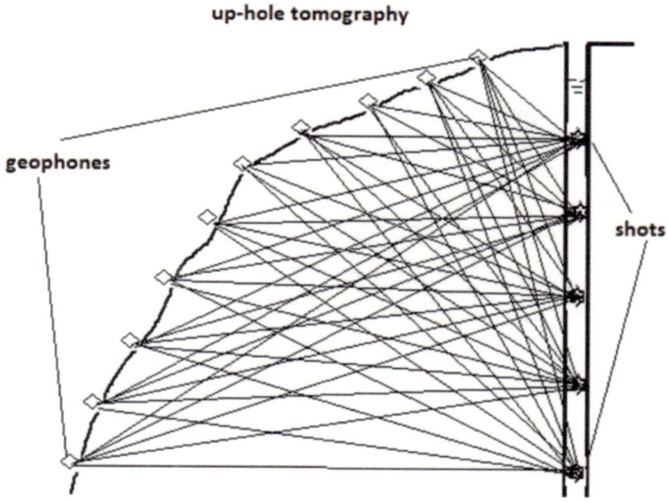

Figure 4.
Acquisition system of the seismic cross-hole tomography.

happen only when the subsoil is homogeneous and isotropic, otherwise, ray paths are curvilinear due to refraction.

3.2 First case study: the ancient walls of the city of Cagliari (Sardinia, Italy)

The study area lies in the northwestern sector of the ancient walls of Cagliari. Apart from the Roman works, of which few traces remain, the original body of the walls, built by the Pisans (Republic of Pisa), dates back to the thirteenth century; in literature, these walls are also referred to as "the Pisan walls" or "the medieval walls." Three centuries later—that is, in in the sixteenth century, after the advent of firearms, in particular the artillery—the Spanish, who at that time dominated Cagliari, decided to modify the defense line incorporating the medieval walls in an embankment coated with limestone blocks; the latter, apart from few modifications and consolidation works, is the current structure of the walls. **Figure 5** shows in detail (the circle) the area in which two up-hole tomographies have been carried out

Figure 5.
The northwestern sector of the Spanish walls. The circle indicates the survey area.

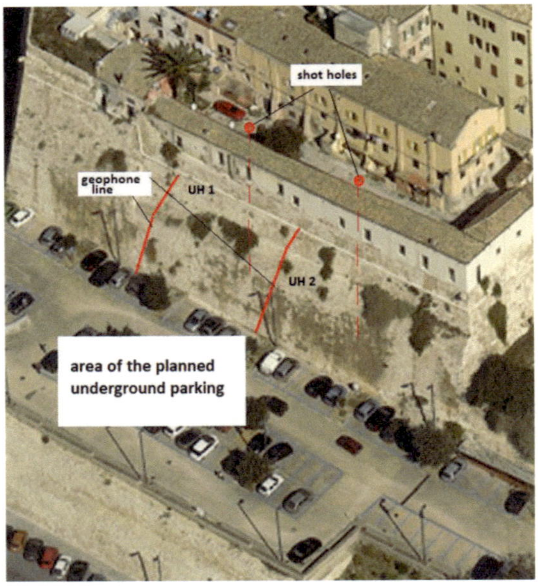

Figure 6.
Position of the two acquisition arrays.

to provide information for the geotechnical modeling of the walls, in view of the construction of an underground car parking, right at the foot of the walls itself.

As for data acquisition geometry, the geophones were placed on the outer surface of the coating at intervals of 1.6 m, and the shots were performed in their respective holes at 1–2 m spacing, starting from the bottom.

Data processing has been performed by means of a software based on a nonlinear optimization technique (ASA, adaptive simulated annealing) [11] that works in terms of modeling, starting from the set of the first arrival times and the spread geometry. **Figure 6** shows the position of the two acquisition arrays, and **Figure 7** shows their schematic view. The results after processing and interpretation are in

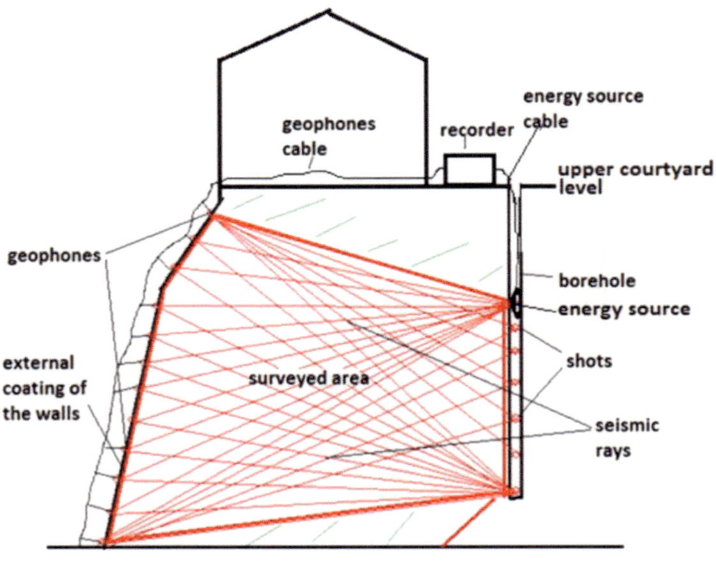

Figure 7.
Schematic view of the up-hole seismic tomographies across the ancient walls.

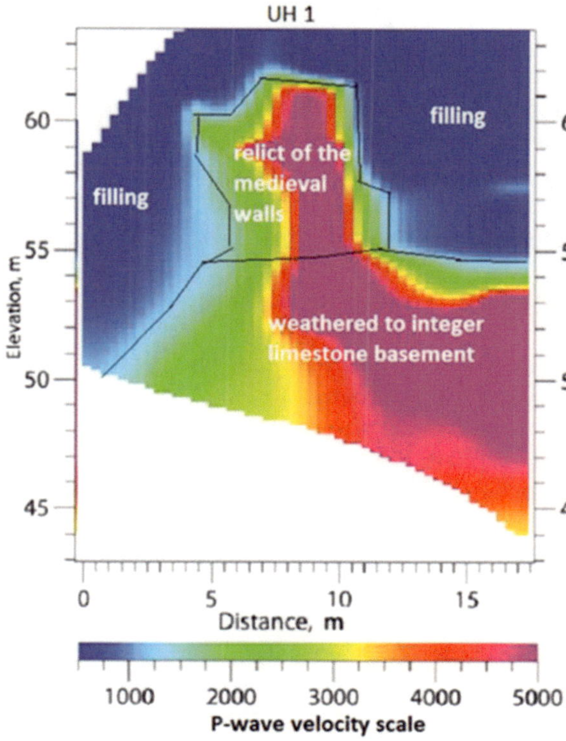

Figure 8.
Up-hole tomography UH1 interpreted.

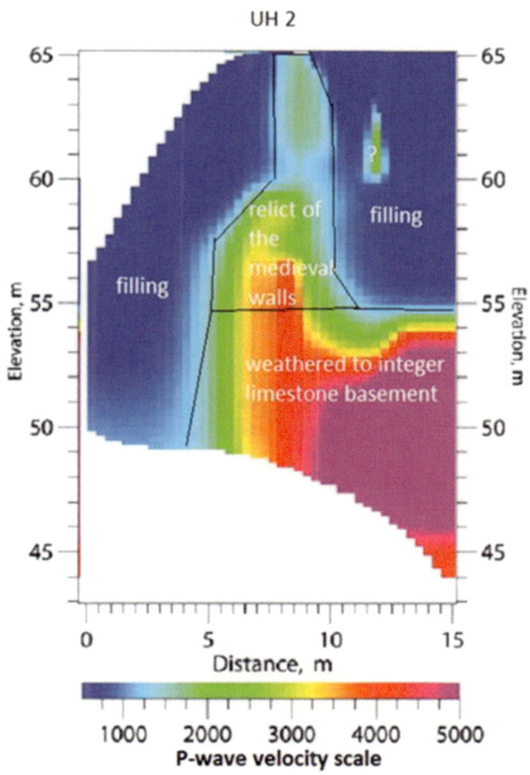

Figure 9.
Up-hole tomography UH2 interpreted.

Figures 8 and **9**. The two tomographies show very similar characteristics, and the ancient medieval wall, the limestone basement, and the filling on both sides of the medieval wall are clearly depicted. On the contrary, the tomographies show no trace of the external coating of the current walls, though its presence is certain. This is a clear sign that thickness of the coating is relatively small and therefore has not been covered satisfactorily by the seismic rays, as deductible from **Figure 7**.

From the two up-hole tomographies and several drillings, the base model for the geotechnical assessment has been compiled as shown in **Figure 10**.

3.3 Second case study: a road embankment along a hillside

Figure 11 shows a satellite view of a road that runs along the side of a hill in Sardinia, Italy. A few years from the construction of the road, after an exceptionally rainy season, progressively pronounced fracture lines, indicated in the figure, appeared on the asphalt. It was the beginning of a landslide that affected the background and the road itself. **Figure 12** shows a detail of a fracture. After 1 month of monitoring, since fracturing—and therefore the landslide—progressed, it was decided to verify the conditions of the subsoil through geophysical techniques. Then, two seismic refraction tomographies perpendicular to the axis of the road, as shown in **Figure 13**, were executed. Data acquisition was carried out with a 48-geophone spread, detector interval of 2.7 m, and same interval for shots (hand hammer with vertical stacking); the processing was performed with the same software employed in the previous case study. The position of the two seismic lines is in **Figure 13**, and the two P-wave velocity sections are in **Figures 14** and **15**.

Tomography SRT1 (**Figure 14**) exhibits a surface zone 5–14 m thick that extends along the whole section. It is composed of the natural unconsolidated overburden and the artificial body of the road embankment, with P-wave velocity in the range 400–800 m/s. The underlying marls are initially much altered for a thickness in the range 2–10 m. Worth of notice is that just under the road, there is a 15–20 m wide zone that deepens for about 20 m, in which, after the demolition of the embankment, water was found. At the base, at depths of 15–20 m from the surface, except

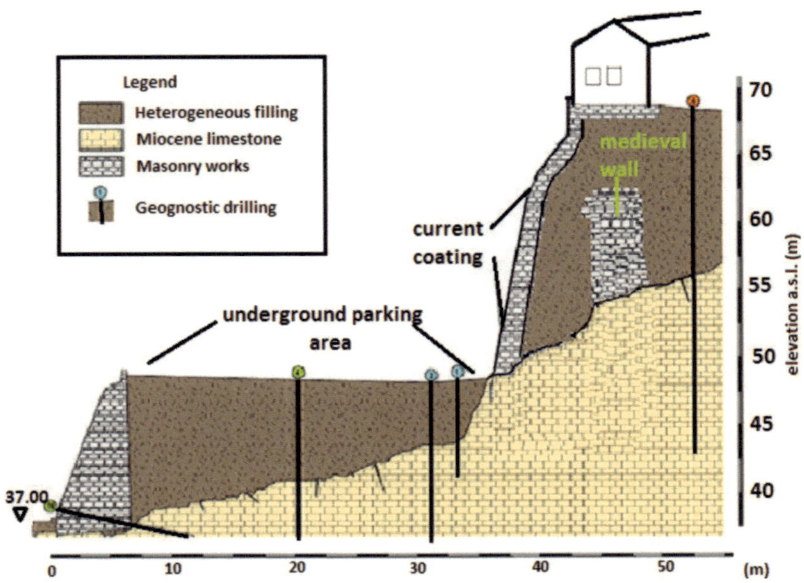

Figure 10.
Geotechnical model deduced from the two up-hole seismic tomographies and four geognostic boreholes indicated in the figure.

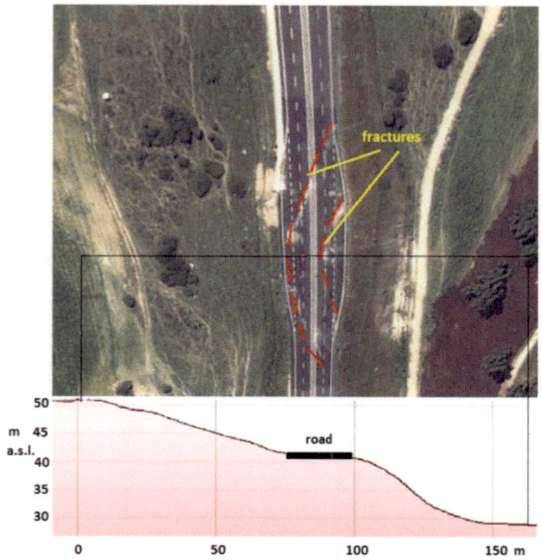

Figure 11.
The stretch of road affected by the landslide phenomenon and the elevation profile.

Figure 12.
Detail of a fracture in the asphalt at the beginning of the landslide phenomena.

Figure 13.
Position of the two seismic refraction tomographies.

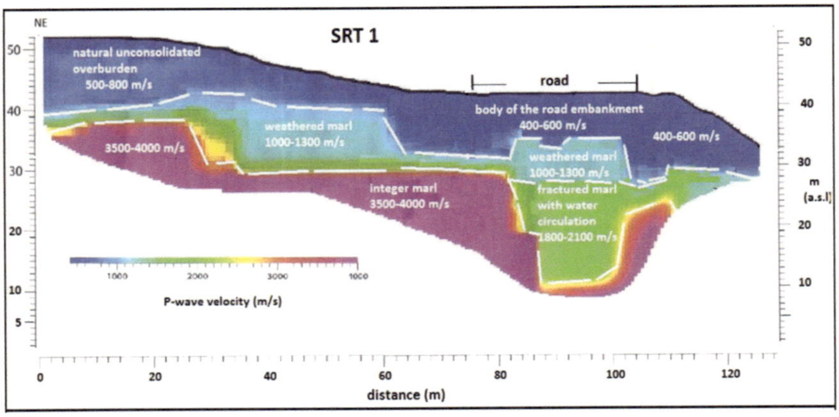

Figure 14.
Seismic refraction tomography SRT 1 interpreted.

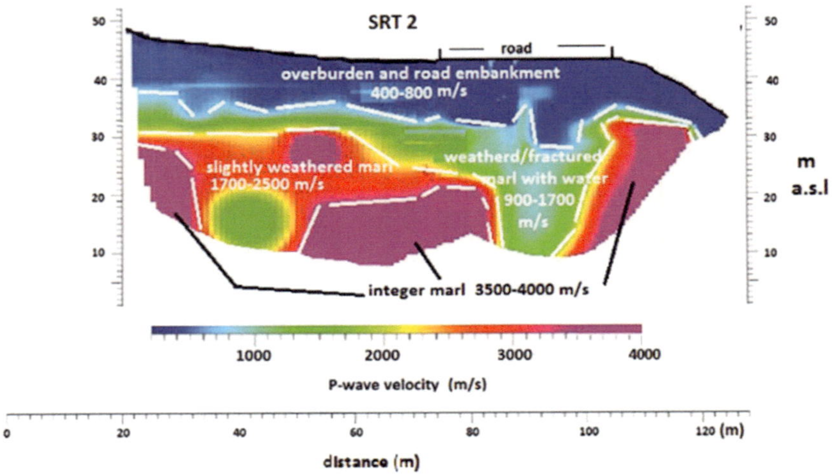

Figure 15.
Seismic refraction tomography SRT 2 interpreted. The SRT 2 is parallel to SRT 1, and the scale indicates the alignment.

in the abovementioned area below the road, there are integer marls. At least three faults can be identified.

The seismic tomography SRT 2 (**Figure 15**) exhibits features similar to those of SRT 1. In this case at least two faults can be identified. The two tomographies will constitute the basis for the road slope stability study.

4. Stability analysis of Spanish walls

4.1 Identification of the geotechnical problem

The geotechnical problem consists of the need of predicting the effects of the excavation in the area adjacent to the bastion and to assess the risk of compromising the stability of the latter. The geognostic and geophysical investigations made it possible to define the internal composition of the bastion and the physical properties of the materials, but they do not allow identifying the actual dimensions of the external coating, that is, the containment wall, and leave uncertain the mechanical properties of the materials, such as the cohesion and the internal friction angle of the Mohr Coulomb model. However, the geognostic and geophysical surveys have made it possible to reconstruct a geomechanical model, whose completion requires a sensitivity analysis of

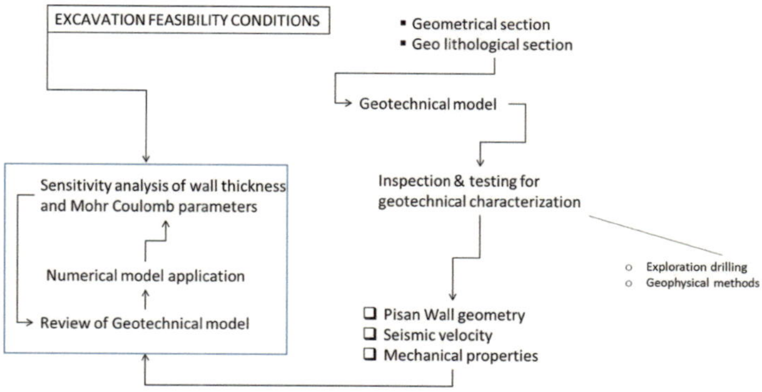

Figure 16.
Case 1: flowchart of iterative analysis modeling—inverse solution.

the results as the geometric and geomechanical characteristics of the external containment wall of the bastion vary. **Figure 16** shows the flowchart of the whole procedure.

4.2 Modeling and results

Calculations were set based on reliable values of geotechnical properties (**Table 1**) called "hypothesis B" and four values (2.5, 2.0, 1.5, and 1.0 m) of the wall thickness. The modeling process allows the calculation of the safety factor (SF) in the three

Material	γ kg m^{-3}	ϕ °	C kPa	σ_tkPa
Rock basement	2500	35	350	130
Spanish coating	2500	35	350	130
Pisan walls	2500	35	350	130
Filling material	1600	35	10	0
Overload	50	—	—	—

Table 1.
Geotechnical properties of the materials for the hypothesis B.

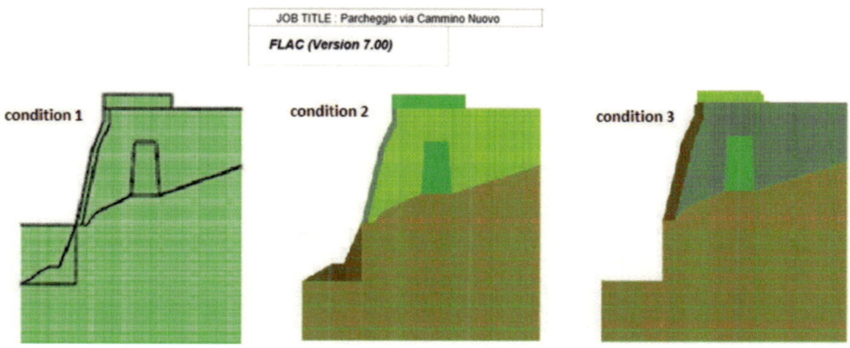

Figure 17.
Left to right: pre-excavation, first excavation, and second excavation.

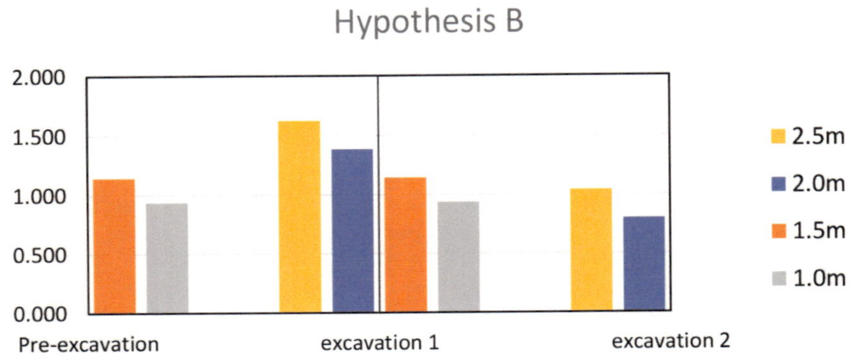

Figure 18.
Results for the hypothesis B: SF versus excavation steps and wall thickness (m).

possible excavation steps, that is, (1) without excavation, (2) excavation of the only loose material, and (3) excavation extended to a portion of the underlying rock (**Figure 17**). The results are shown in histograms of **Figure 18**.

The hypothesis B has been modified in A and C that differ for the cohesion and the tensile strength. The results obtained are illustrated in the diagrams of **Figure 19**.

4.3 Feasibility of excavation

The results obtained show that:

- Changes in the mechanical properties result in unimportant changes in the SF. For variations of cohesion in a range of ±20% and of the tensile strength of ±40%, the corresponding SF values vary by only ±5%.

- The realization of the first excavation, that is, the removal of the ground in front of the wall (**Figure 17**), does not cause significant SF variations that can

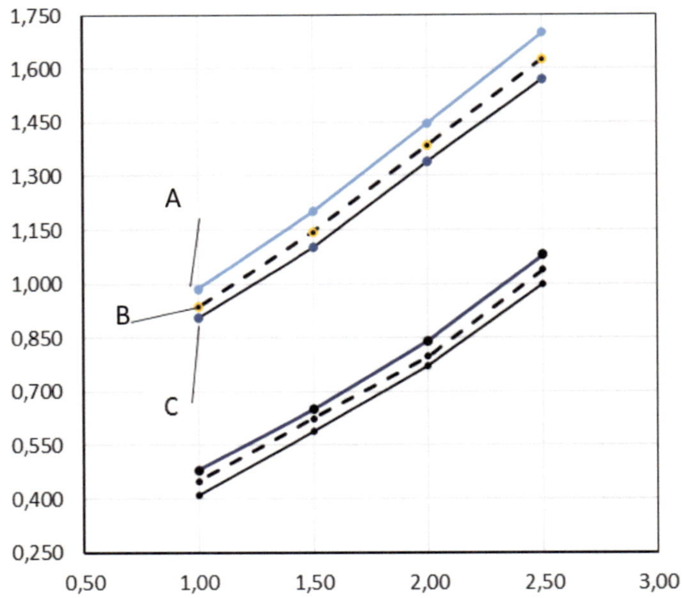

Figure 19.
SF versus wall thickness (m) for A, B, and C hypotheses of mechanical properties.

be considered constant (see **Figure 18**) for any wall thickness. The subsequent excavation n.2, corresponding to the removal of a rock portion, determines reductions in the SF (see diagrams in **Figures 18** and **19**) between 30% and 55%. As less thick is the wall as smaller is the SF (see **Figure 20**).

- The possibility and the ways of carrying out the excavation in front of the wall are linked to the verification of its thickness.

In summary, in all the cases examined, the execution of the excavation must always be preceded by retaining structures; these must concern only the base of the wall, if only the first excavation will be carried out, or even the rock wall, if the second excavation will be carried out. The retaining structures design will have to consider the effective thickness of the wall and will have most important and binding extension, the lower the measured thickness will be (**Figure 20**).

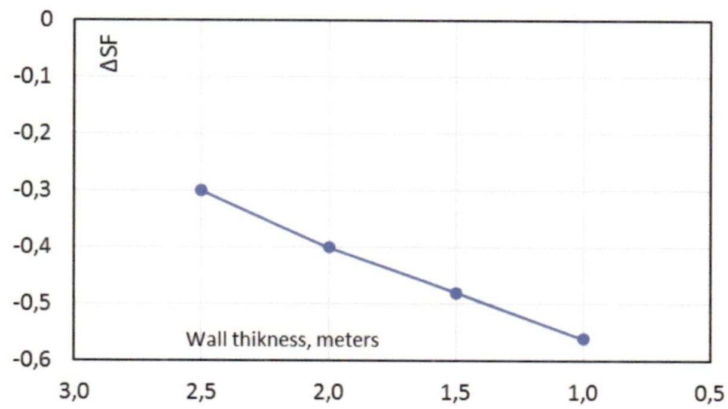

Figure 20.
SF reduction for the second excavation versus wall thickness (meters).

5. Road slope stability

5.1 Identification of the geotechnical problem and modeling results

In the second case, the geotechnical problem is represented by a slope instability, apparently unjustified even in the case of poor geotechnical properties of the materials but occasionally worsened by groundwater from a nearby stream [12, 13].

Rock type	Density, kg m^{-3}	Young's modulus, MPa	Poisson's ratio	Cohesion, kPa	Friction angle	Tensile strength, kPa
Compact marl	2300	500	0.25	200	35°	50
Altered marl	2100	250	0.25	0–5 to 10–15	30°	0
Arenaceous rock	2000	100	0.30	10–15	30°	0
Fill road material	1600	50	0.30	10–15	25°	0

Table 2.
Geotechnical properties of the materials.

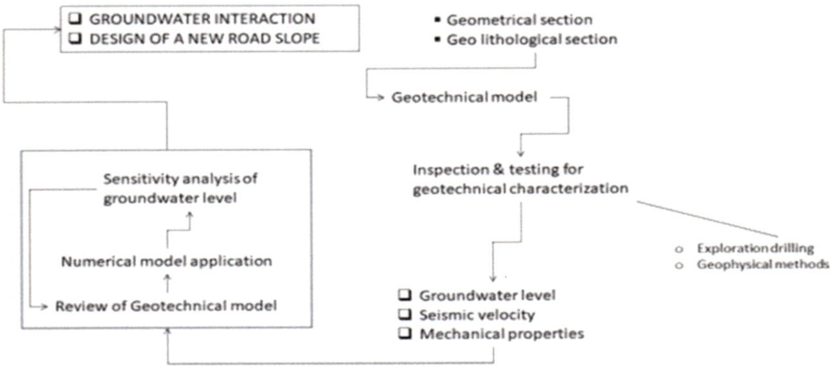

Figure 21.
Case 2: flowchart of iterative analysis modeling—Sensitivity calculations.

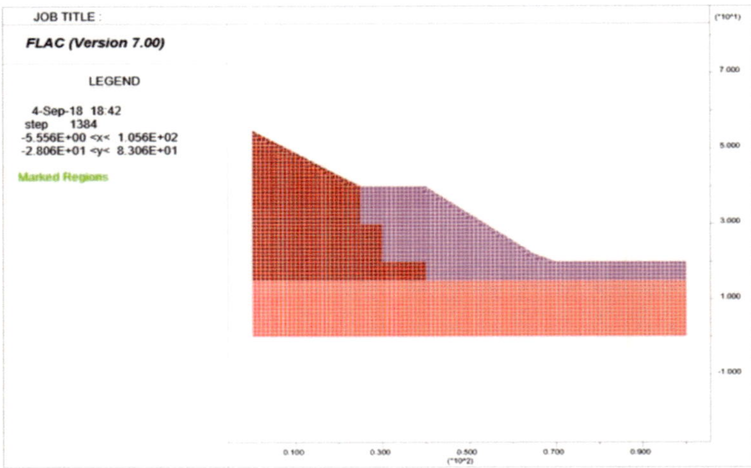

Figure 22.
Numerical model of the slope.

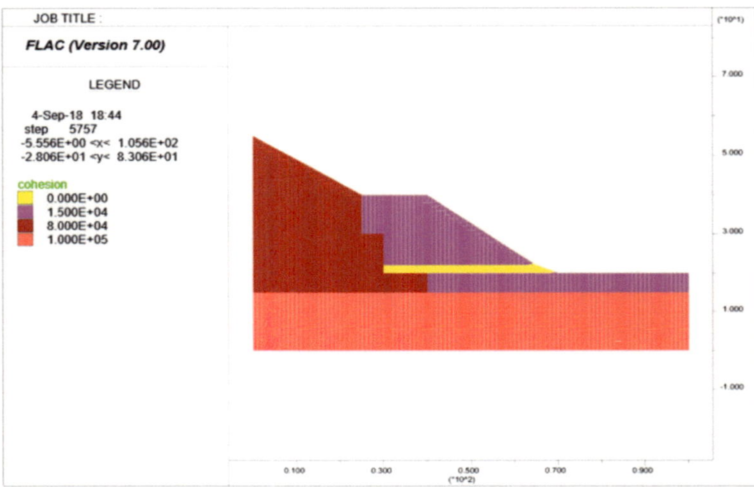

Figure 23.
Cohesion of the different materials.

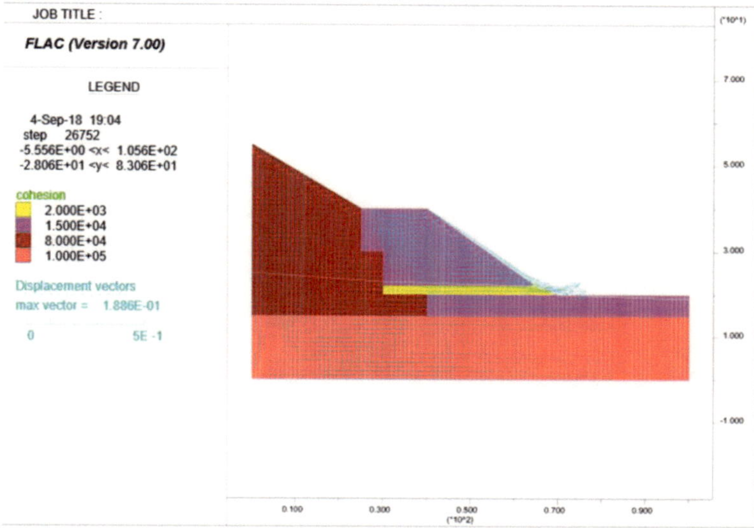

Figure 24.
Displacement vectors of the slope instability.

In accordance with the results of exploration drilling and geophysical survey (see geotechnical properties in **Table 2** and flowchart of **Figure 21**), a geotechnical model was reconstructed whose solution was obtained with numerical methods and the application of the 2D FLAC software [2] (see **Figures 22–24**).

The numerical model was corrected according to an iterative procedure to reconstruct the instability mechanism observed in situ. The next step was to highlight the effects of a growing groundwater. The datum to which the simulation referred was the presence or not of the groundwater level (GL or not GL in the **Figure 25**).

The results obtained are highlighted in the diagrams of **Figure 25**, which show the trend of the SF versus the cohesion of the altered marls, for two cohesion conditions (15 and 10 kPa) of intact marls. The diagrams show that to verify instability (SF ≤ 1), two conditions are necessary: (i) the presence of a water table under the

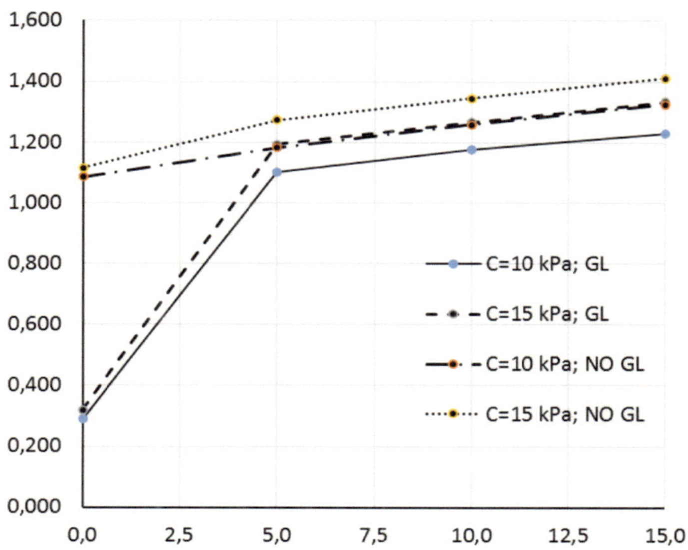

Figure 25.
SF versus altered marlstone's cohesion (kPa) for two conditions of intact marls (C = 10 and 15 kPa) and groundwater level presence (GL) and absent (NO GL).

roadway and (ii) the circulation of water that causes the reduction of cohesion in the altered marls below the value of 5 kPa.

5.2 Road slope instability

The results show that the groundwater presence (GL) and a consequent reduction in cohesion of the altered marl horizon determine a drastic reduction of the safety factor below the unit value, as illustrated in **Figure 24** (displacement vectors) and in the diagrams of **Figure 25**.

6. Conclusions

Some aspects that are not directly investigated characterize the situations examined. The first one concerns the materials constituting a wall system built in the fifteenth and sixteenth centuries, completely integrated into the city center; this wall system is bordered by an area in which an underground car parking is planned and the excavation could create dangerous conditions for the wall system itself. In the second case, problems arise from the effects of groundwater in the alluvial soil covering a hill, on the flank of which runs a road. Here, in the case of intense meteoric conditions, the groundwater reaches the foot of the slope and determines its instability, by rotational kinematics, in an unexpected way.

In both cases, traditional geotechnical investigations do not solve the problems. Conversely, geophysical surveys integrate the knowledge framework and provide the fundamental elements for the development of back or sensitivity analysis, which can be performed with numerical methods, based on reliable geomechanical models.

Based on the results obtained, it was possible to conclude that:

- In the first case, it should be noted that to carry out the excavation, preliminary retaining structures (cables, micro piles, etc.) are necessary, whose designing must take into account both the effective wall thickness and the extent of the excavation, whether it will concern only the loose material or even the rock below.

- In the second case, it should be noted that drainage and waterproofing of the slope toward the surrounding land will be necessary.

Conflict of interest

The authors of this chapter declare that in its publication, there is no reason, even potential, of conflict of interest.

Author details

Roberto Balia* and Pier Paolo Manca
Dipartimento di Ingegneria Civile, Ambientale e Architettura, Università di Cagliari, Italy

*Address all correspondence to: balia@unica.it

References

[1] Hudson JA. Rock engineering case histories: Key factors, mechanisms and problems. In: Särkkä, Eloranta, editors. Rock Mechanics—A challenge for society. Proceedings of ISRM Regional Symposium EUROC2001; Espoo, Finland; 4-7 June 2001; Rotterdam: Balkema; 2001

[2] Detournay C, Hart R. FLAC and numerical modelling in geomechanics. In: Proceedings of the International FLAC Symposium on Numerical Modelling in Geomechanics, Minneapolis; Rotterdam: Balkema; 1999

[3] Jing L. Mint: A review of techniques, advances and outstanding issues in numerical modelling for rock mechanics and engineering. International Journal of Rock Mechanics and Mining Sciences. 2003;**40**:283-353

[4] Van Westen CJ et al. Mint: Prediction of the occurrence of slope instability phenomena through GIS-based hazard zonation. Geologische Rundschau. 1997;**86**:404-414

[5] Borja RI. Mint: WJA. Continuum deformation and stability analyses of a steep hillside slope under rainfall infiltration. Acta Geotechnica. 2010;**5**:1-14

[6] Dobrin MB, Savit CH. Introduction to Geophysical Prospecting. 4th ed. New York: McGraw-Hill; 1988. ISBN: 0070171963; 007100404

[7] Foti S, Lai CG, Rix GJ, Strobbia C. Surface Wave Methods for Near-Surface Site Characterization. London: CRC Press; 2017

[8] Palmer D. The Generalized Reciprocal Method of Seismic Refraction Interpretation. Tulsa, Oklahoma: Society of Exploration Geophysicists; 1980. ISBN (Print): 978-0-931830-14-3

[9] Peterson JE, Paulsson BN, McEvilly TV. Mint: Applications of algebraic reconstruction techniques to crosshole seismic data. Geophysics. 1985;**50**:1566-1580

[10] White DJ. Mint: Two-dimensional seismic refraction tomography. Geophysical Journal. 1989;**97**:223-245

[11] Ingber L. Mint: Adaptive simulated annealing (ASA): Lessons learned. Control and Cybernetics. 1996;**25**(1):33-54

[12] Montgomery DR. Mint: Road surface drainage, channel initiation, and slope instability. Water Resources Research. 1994;**30**:1925-1932

[13] Fourie AB. Predicting rainfall-induced slope instability. Proceedings of the Institution of Civil Engineers: Geotechnical Engineering. 1996;**119**(4):211-218

Chapter 7

Electrical Resistivity Tomography: A Subsurface-Imaging Technique

Bing Zhou

Abstract

Electrical resistivity tomography (ERT) is a popular geophysical subsurface-imaging technique and widely applied to mineral prospecting, hydrological exploration, environmental investigation and civil engineering, as well as archaeological mapping. This chapter offers an overall review of technical aspects of ERT, which includes the fundamental theory of direct-current (DC) resistivity exploration, electrode arrays for data acquisition, numerical modelling methods and tomographic inversion algorithms. The section of fundamental theory shows basic formulae and principle of DC resistivity exploration. The section of electrode arrays summarises the previous study on all traditional-electrode arrays and recommends 4 electrode arrays for data acquisition of surface ERT and 3 electrode arrays for cross-hole ERT. The section of numerical modelling demonstrates an advanced version of finite-element method, called Gaussian quadrature grid approach, which is advantageous to a numerical simulation of ERT for complex geological models. The section of tomographic inversion presents the generalised standard conjugate gradient algorithms for both the l_1- and l_2-normed inversions. After that, some synthetic and real imaging examples are given to show the near-surface imaging capabilities of ERT.

Keywords: resistivity, electrical current, geotomography, numerical modelling, subsurface imaging

1. Introduction

Direct-current (DC) resistivity exploration is a traditional geophysical method. It employs two electrodes to inject electric current into the ground and other two electrodes to measure the electric potential difference. The measurements are often carried out along a line or in an area on the earth surface, and then the observed potential differences are converted into sounding curves or pseudo-sections of apparent resistivities, which indicate the resistivity changes of subsurface rocks. Analyses of these data enable us to find the underground resistivity anomalies or outline the subsurface geological structure. With development of computer technology and numerical computational techniques, accurate numerical simulations of subsurface electrical field and acquiring a large amount of data in fields become possible [1–3], so that the traditional DC resistivity exploration was developed to a computerised geotomography technique, called electrical resistivity tomography (ERT), which employs a multielectrode equipment or system to automatically acquire a large number of data [4, 5] and applies a computer software to the reconstruction of subsurface resistivity structure with the observed data [6–10]. Due to its conceptual simplicity, low equipment cost and ease of use, ERT is now widely applied in mineral exploration, civil engineering, hydrological prospecting and environmental investigations, as well as archaeological mapping [11]. This chapter provides an overall review of ERT techniques, which consists of four sections: (1) fundamental theory, (2) electrode arrays, (3) numerical modelling and (4) tomographic inversion. In each section, diagrams and formulations are used to illustrate basic concepts and principles of ERT techniques. Some synthetic experiments and practical imaging applications are also given to show the imaging capability of ERT.

2. Fundamental theory

According to the continuity of electrical current, the following integral equation is satisfied at any point in a conductive medium:

$$\oint\!\!\!\oint_{\Gamma} \boldsymbol{J} \cdot d\Gamma = -I, \tag{1}$$

where Γ is a full or half spherical surface that encloses an electrode that injects electric current $\boldsymbol{J}$ with a magnitude of I (**Figure 1**). According to Ohm's law:

$$\boldsymbol{J} = \sigma \boldsymbol{E} = -\sigma \nabla U, \tag{2}$$

and the property of the delta-function $\delta(\boldsymbol{x}\text{-}\boldsymbol{x}_s)$, Eq. (1) gives

$$-\oint\!\!\!\oint_{\Gamma} \sigma \nabla U \cdot d\Gamma = -I \iiint_{\Omega} \delta(\boldsymbol{x} - \boldsymbol{x}_s) d\Omega \tag{3}$$

Here σ is the conductivity of medium, U is the electric potential, Ω represents the volume of the spherical surface Γ, and $\boldsymbol{x}_s = (x_s, y_s, z_s)$ is the location of the current electrode. Applying the divergence theorem to Eq. (3) leads to

$$\iiint_{\Omega} [\nabla \cdot (\sigma \nabla U) + I\delta(\boldsymbol{x} - \boldsymbol{x}_s)] = 0. \tag{4}$$

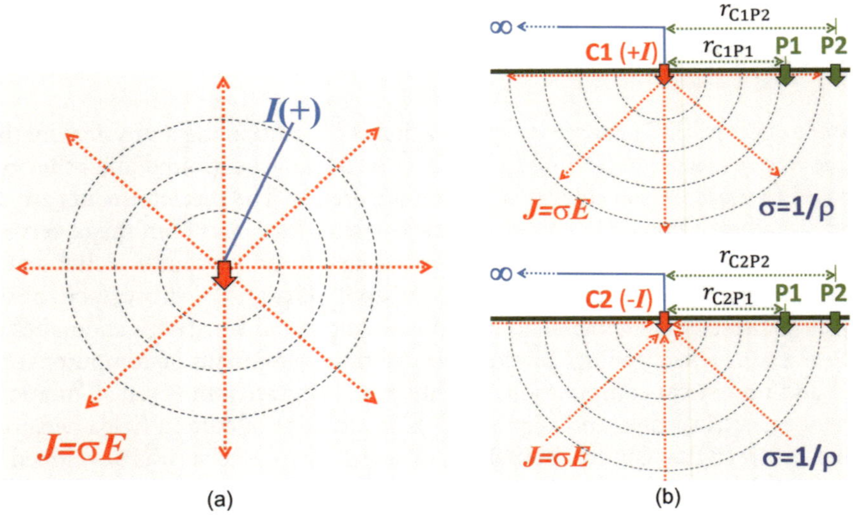

Figure 1.
Electric current density J around an electrode in (a) a full space and (b) a half-space.

Note that Eq. (4) satisfies everywhere in a medium, so that the following governing equation of electric field is obtained:

$$\nabla.(\sigma \nabla U) = -I\delta(\boldsymbol{x} - \boldsymbol{x}_s), .\boldsymbol{x} \in \Omega. \quad (5)$$

In general, the conductivity $\sigma(\boldsymbol{x}) = \sigma(x,y,z)$ changes with three coordinates so that it defines a 3D geological model, and the electric potential $U=U(x,y,z)$ becomes a 3D function of a given current injection $I\delta(\boldsymbol{x} - \boldsymbol{x}_s)$ and conductivity $\sigma(\boldsymbol{x})$ (**Figure 2a**). If the conductivity is a constant with the y-coordinate, i.e. $\sigma(\boldsymbol{x}) = \sigma(x,z)$, it defines a 2D geological model (**Figure 2b**), but the electric potential $U(x,y,z)$ is still a 3D function of the coordinates due to the point current injection $I\delta(\boldsymbol{x} - \boldsymbol{x}_s)$. Applying cosine Fourier transform $F_c\{\cdot\}$ to the y-coordinate of Eq. (5), the governing equation is changed into

$$\nabla.\left(\sigma \nabla \widetilde{U}\right) + k_y^2 \widetilde{U} = -I\delta(\boldsymbol{x} - \boldsymbol{x}_s)/2, \boldsymbol{x} \in \Omega, \quad (6)$$

which is often named 2.5D governing equation. Here k_y is the wavenumber, and $\widetilde{U}(x, k_y, z) = F_c\{U(x,y,z)\}$ becomes the spectrum of electric potential in the wavenumber domain. The coordinate vector $\boldsymbol{x} = (x,z)$ and gradient $\nabla = (\partial_x, \partial_z)$ are the 2D versions. If a geological model $\sigma(x,y,z)$ or $\sigma(x,z)$ and current injection $I\delta(\boldsymbol{x} - \boldsymbol{x}_s)$ are given, solving the governing equation (5) or (6), one obtains electrical potential $U(x,y,z)$ for a 3D conductivity model or its spectrum $\widetilde{U}(x, k_y, z)$ for a 2D conductivity model and then performs the inverse cosine Fourier transform $U(x, y, z) = F_c^{-1}\left\{\widetilde{U}(x, k_y, z)\right\}$ to obtain electric potential $U(x,y,z)$. These computations are called *forward modelling*. In theoretical computation or numerical forward modelling, Green's function G or $\widetilde{G}$—the electric potential response to a unit current injection ($I = 1$ Am)—is often applied, so as to remove the magnitude of electric current I in Eqs. (5) and (6). One may focus on the computation of Green's function, and then the electric potential U or $\widetilde{U}$ is computed by the multiplication of Green's function G with the magnitude of practical electric current I, i.e. $U = I{\cdot}G$.

The simplest geological model is a homogenous half-space. Applying Eq. (3) (equivalent to Eq. (5)) to a constant medium, the surface integral is calculated by

$$\oiint_\Gamma \sigma \nabla U \cdot d\Gamma = \sigma \nabla U\left(2\pi r^2\right) = -I, \quad (7)$$

which gives the electric potential at distance r from a current source:

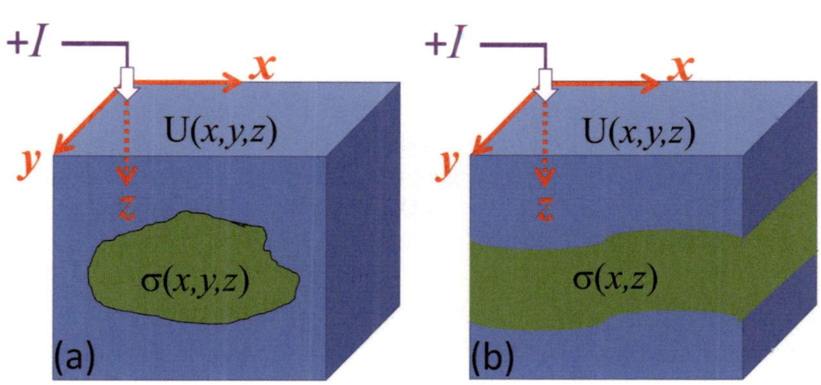

Figure 2.
Sketch of (a) a 2D and (b) a 3D geological model defined by conductivity σ(x,z) and σ(x,y,z), respectively.

$$U = \frac{I}{2\pi\sigma r} = \frac{\rho I}{2\pi r}, \tag{8}$$

where $\rho = 1/\sigma$ is resistivity. In practice, to inject electric current into the ground, a pair of current electrodes $C1$ and $C2$ must be employed; one is positive ($+I$) and another is negative ($-I$). Thus, the electric potential at a point P with a pair of current electrodes is calculated by.

$$U = \frac{\rho I}{2\pi}\left(\frac{1}{r_{C1P}} - \frac{1}{r_{C2P}}\right). \tag{9}$$

Here, r_{C1P} and r_{C2P} are distances of the observed point P to two current electrodes $C1$ and $C2$. To measure the potential on the earth surface, a pair of potential electrodes $P1$ and $P2$ is also required (**Figure 1**). According to Eq. (9), one has the following expression for the electric potential difference between two potential electrodes:

$$\Delta U = \frac{\rho I}{2\pi}\left[\left(\frac{1}{r_{C1P1}} - \frac{1}{r_{C2P1}}\right) - \left(\frac{1}{r_{C1P2}} - \frac{1}{r_{C2P2}}\right)\right], \tag{10}$$

from which *apparent resistivity* ρ_a is defined as follows:

$$\rho_a = K\frac{\Delta U}{I} = K\Delta G, \tag{11}$$

where $\Delta G = \Delta U/I$ and K is called geometry factor of electrode array given by

$$K = 2\pi\left[\left(\frac{1}{r_{C1P1}} - \frac{1}{r_{C2P1}}\right) - \left(\frac{1}{r_{C1P2}} - \frac{1}{r_{C2P2}}\right)\right]^{-1}, \tag{12}$$

which depends on the positions of four electrodes. Different layouts of four electrodes have variable geometry factors and are often called *electrode arrays*. In the traditional electrode arrays, $C2$ and $P2$ may be set up very far away from $C1$ and $P1$, so that $C2$ and $P2$ are treated as remote electrodes in theory, and distances r_{C1P2}, r_{C2P2}, r_{C2P1}, and r_{C2P2} are supposed to be infinite (∞) in Eq. (12). In these cases, one can find that geometry factors are still applicable for these electrode arrays that involve one or two remote electrodes, which are named pole-pole, pole-dipole and dipole-pole arrays.

If subsurface resistivity is homogenous (ρ_0), Eq. (11) shows that no matter which electrode array is used, apparent resistivity is constant ($\rho_a = \rho_0$). Otherwise, ρ_a indicates resistivity variation of the underground. For a certain range of apparent resistivity ρ_a, Eq. (11) also reveals that the geometry factor is inversely proportional to the potential difference $\Delta G = \Delta U/I$. It implies that a big value of geometry factor K will cause a small reading of the electric potential difference ΔG in fields. Such a small reading is easily obscured or contaminated by background noise. For a good data quality, one should avoid very big values of geometry factors in data acquisition for ERT. Therefore, Eqs. (11) and (12) are the fundament formulae of the traditional DC electrical resistivity exploration.

3. Electrode arrays

In order to obtain apparent resistivity in fields, many electrode arrays were developed in the traditional DC resistivity exploration. In principle, ERT requires a high data density and good coverage of the earth surface for high-resolution images of subsurface targets. Dahlin and Zhou [11] carried out synthetic experiments of ERT using 10 electric arrays and compared their imaging results for four geological models: a buried channel, a narrow dike, dipping blocks and waste ponds. They demonstrated that two three-electrode arrays (pole-dipole and dipole-pole) and three four-electrode arrays (dipole–dipole, Schlumberger and gradient arrays) produce satisfactory images of the subsurface targets. However, due to the use of remote electrodes, the three-electrode arrays are rarely applied for ERT in practice; thus, the four-electrode arrays become popular. Particularly, gradient array [12] is well suited for multichannel data acquisition and can significantly increase the speed of data acquisition in the field, and at the same time, it gives higher data density and lower sensitivity to noise than dipole-dipole array. **Figure 3** shows three four-electrode arrays (upper row) and their pseudo-sections of data points (x_{ρ_a}, z_{ρ_a}) (middle row) and their geometry factors (bottom row) for a layout of total 81 electrodes on the earth surface. **Figure 3** shows that dipole–dipole array has the biggest range of geometry factor among these three-electrode arrays, and gradient array performs the smallest range, so that it is well suitable for data acquisition with a high data density and good data quality. Zhou and Dahlin [13] based on many sets of field data obtained with different electrode arrays and found that the measured potential errors depend on the reading of potential difference, $\Delta G = \Delta U/I$, and the measured datum in field may be expressed by

$$\Delta G^* = \Delta G\left[1 + R\left(\frac{\alpha}{\Delta G}\right)^{\beta}/100\right], \tag{13}$$

where ΔG^* denotes a noise-contaminated datum, R is a random number in (−1,1) and α and β are constants. Due to having a big range of geometry factor (see **Figure 3**), the dipole-dipole array is more sensitive to noise than Schlumberger and gradient arrays, so that one must pay much attention to the noise contamination when conducting dipole-dipole measurements [13]. According to the relationship between the geometry factor K and reading of potential difference ΔG, Eq. (13) may be applied to predications of noise-contaminated data for different electrode arrays [11].

Reviewing **Figure 3**, one may find that gradient array only uses the potential electrodes between $C1$ and $C2$ to measure the potential differences. It does not do the measurements at the potential electrodes outside of $C1$ and $C2$. It means that gradient array actually misses the outside data for every pair of current electrodes. To obtain a better data coverage and complete subsurface information, the outside

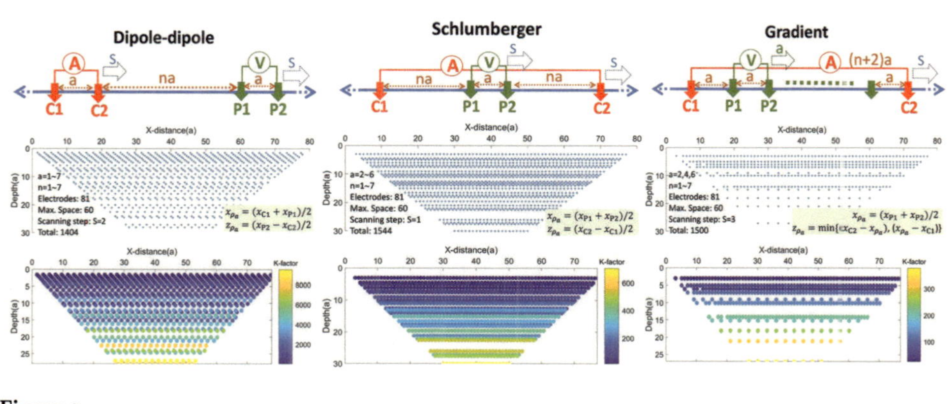

Figure 3.
Dipole-dipole, Schlumberger and gradient arrays (upper row) for ERT data acquisition and examples of their pseudo-sections of data points (x_{ρ_a}, z_{ρ_a}) (middle row) and geometry factors (bottom row).

potential electrodes of each pair of *C*1 and *C*2 should be employed in the gradient measurements. These additional data apparently complement to common gradient measurements and may improve ERT imaging. Accordingly, a new electrode array is naturally formed and called *full-range gradient array* shown in **Figure 4a**. The difference from the common gradient array is that the new electrode array uses not only the potential electrodes between *C*1 and *C*2 but also the outside potential electrodes of every pair of *C*1 and *C*2, if their geometry factors fall in a reasonable range for the data acquisition. **Figure 4b** and **c** shows the pseudo-section points and geometry factors for a layout of total 81 electrodes. Comparing with **Figure 3**, the full-range gradient array does improve the data coverage, and its geometry factors are controlled in a reasonable range.

If there are boreholes in a field, cross-hole ERT may be carried out to image the geological structure between the boreholes. Zhou and Greenhalgh [14] investigated all possible electrode arrays for cross-hole ERT data acquisition and found that the electrode arrays of pole-pole (A-M), pole-bipole (A-MN), bipole-pole (AB-M), and bipole-bipole (AM-BN) with their multi-spacing cross-hole profiling and scanning surveys are useful for cross-hole ERT. Here the capital letters A and B stand for two current electrodes, and M and N denote two potential electrodes. These cross-hole electrode arrays are shown in **Figure 5** with their sensitivity functions in backgrounds [15]. They also found that the electrode arrays which have either both current electrodes or both potential electrodes in the same borehole, e.g. A-MN, AB-M and AB-MN, have a singularity problem in data acquisition (geometry factor goes to infinite so that apparent resistivity and pseudo-reaction are not applicable), namely, zero readings of the potential or potential difference in cross-hole measurements, so that the potential data are easily obscured by background noise and their images are inferior to those from other

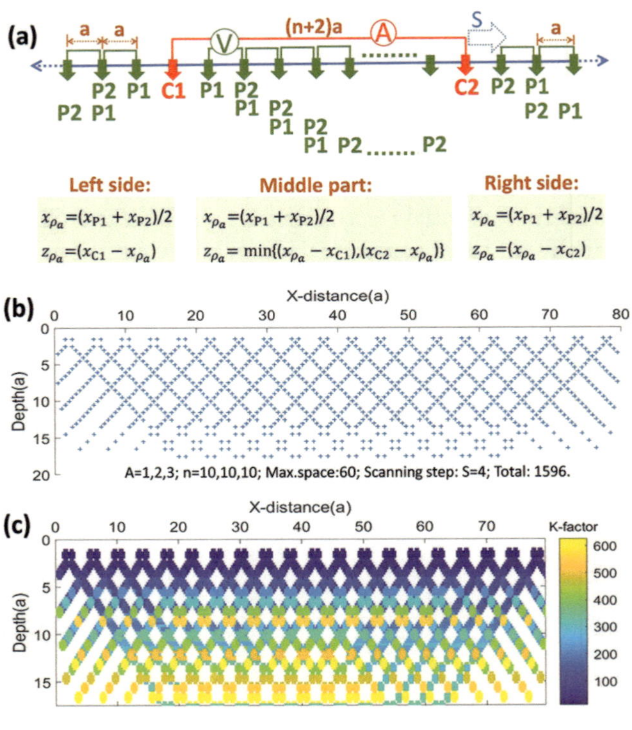

Figure 4.
Full-range gradient array (a) for ERT data acquisition and examples of its pseudo-section of data points (x_{ρ_a}, z_{ρ_a}) *(b) and geometry factors (c) with a layout of total 81 electrodes.*

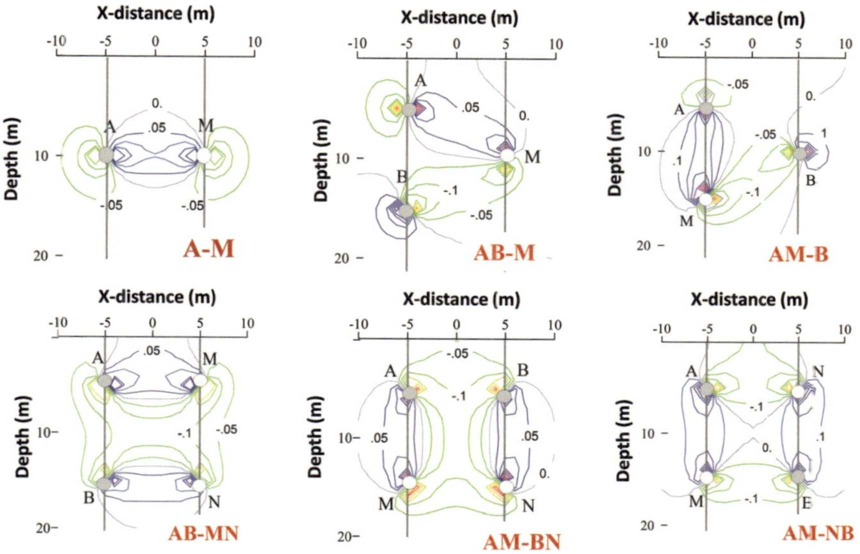

Figure 5.
Electrode arrays for cross-hole ERT data acquisition. A and B stand for two current electrodes. M and N denote two potential electrodes. The background contours are the sensitivity functions of the electrode array.

cross-hole electrode arrays. The data having the singularity problem may be predicated by zero values of the inverse geometry factors, which should be avoided in cross-hole data acquisition. Therefore, A-M, AM-B, and AM-BN with multi-spaces are recommended for cross-hole ERT.

4. Numerical modelling

To compute theoretical electric potential $U(x,y,z)$ in a complex 2D or 3D geological model, one has to solve the governing equation (6) or (5) using a numerical approach and a computer. For computational efficiency, an underground artificial boundary Γ_0 is required to truncate an infinite geological model, so that the boundary condition on Γ_0 must be introduced to the governing equations. Applying Eq. (7) to the artificial boundary Γ_0 and zero normal component ($\mathbf{J}\cdot\mathbf{n} = 0$) of electric current density $\mathbf{J}$ on the earth surface, one may find that the electric potential U holds

$$\sigma \nabla U \cdot \mathbf{n} + \nu U = 0, \tag{14}$$

where $\mathbf{n}$ is the normal vector of the artificial boundary Γ_0 and ν is calculated by $\nu = (\mathbf{r}\cdot\mathbf{n})/r^2$. Eq. (14) is the 3D boundary condition on Γ_0. Similarly, one may find the boundary condition for a 2D geological model [16]:

$$\sigma \nabla \widetilde{U} \cdot \mathbf{n} + \nu \widetilde{U} = 0. \tag{15}$$

Adding the artificial boundary conditions Eqs. (14) and (15) to Eq. (5) and (6), numerical modelling becomes to solve the following definite governing equation:

$$3\text{D}: \begin{cases} \nabla \cdot (\sigma \nabla G) = -\delta(\boldsymbol{x} - \boldsymbol{x}_s), \boldsymbol{x} \in \Omega, \\ \sigma \nabla G \cdot \mathbf{n} + v G = 0, \ \boldsymbol{x} \in \Gamma_0 \end{cases} \tag{16}$$

or

$$2\text{D}:\begin{cases}\nabla\cdot\left(\sigma\nabla\tilde{\text{G}}\right)+\text{k}_y^2\tilde{\text{G}}=-\delta(\boldsymbol{x}-\boldsymbol{x}_s)/2, \boldsymbol{x}\in\Omega,\\ \sigma\nabla\tilde{\text{G}}\cdot\mathbf{n}+\upsilon\tilde{\text{G}}=0,\ \ \boldsymbol{x}\in\Gamma_0.\end{cases} \tag{17}$$

The numerical approach to the definite governing equations is called *numerical modelling*. The most popular numerical approaches are finite-difference and finite-element methods. The former is simple and straightforward, due to directly applying a finite-difference formula to the derivatives of the gradient ∇, so that the definite governing equations at all points in the model domain Ω and on the artificial boundary Γ_0 are discretised and assembled into a linear equation system [16–20]. The latter converts the definite governing equation into an integral equation by weighting residual principle or variational principle [21] and then carries out numerical volume integration to produce a linear equation system [22–26]. Therefore, no matter which approach is applied, the numerical modelling becomes to solve a linear equation system. For an instant, the following paragraphs briefly show an advanced 2D version of the finite-element method, called Gaussian quadrature grid (GQG) approach, because of its advantages over finite-difference method and other schemes of finite-element approaches, e.g. high accuracy and easy implementation of discretisation of a geological model having arbitrary free-surface topography. For a 3D numerical modelling, one may follow the 2D procedures.

Weighting residual principle is to calculate the following integral of the governing equation in Eq. (17):

$$\iint_{\Omega}\text{W}\left[\nabla\cdot\left(\sigma\nabla\tilde{\text{G}}\right)+k_y^2\tilde{\text{G}}+\delta(\boldsymbol{x}-\boldsymbol{x}_s)/2\right]d\Omega, \tag{18}$$

where W is an arbitrary weighting function. Applying the divergence theorem to the above and then submitting the artificial boundary condition yields

$$\iint_{\Omega}\left(\sigma\nabla W\cdot\nabla\tilde{\text{G}}+\sigma k_y^2 W\tilde{\text{G}}\right)d\Omega+\int_{\Gamma_0}\nu W\tilde{\text{G}}d\Gamma=\text{W}(\boldsymbol{x}_s)/2. \tag{19}$$

In order to calculate the integrals, the model domain Ω is divided into a set of the no-overlap subdomains $\{\Omega_e, e = 1,2,\ldots,N_e\}$ that matches the free-surface topography and subsurface interfaces. In each subdomain, Gaussian abscissae $\{(x_\alpha, z_\beta), \alpha,\beta = 1,2,\ldots,N_G\}$ are employed (see examples shown in **Figure 6**) to discretise the subdomain and then Lagrange interpolation:

$$\tilde{\text{G}}=\sum_{i,j=1}^{N_G}l_i(x), l_j(z)\tilde{\text{G}}_{ij}^{(e)}, \tag{20}$$

and Gaussian weights $\{w_{\alpha\beta}\}$ are applied to calculation of the submain integrals. Consequently, Eq. (19) becomes

$$\begin{aligned}\sum\nolimits_e\Big\{&\sum\nolimits_{\alpha,\beta,i,j}w_{\alpha\beta}\sigma_{\alpha\beta}\left[l_i'(x_\alpha)\delta_{j\beta}\partial_x\text{W}_{\alpha\beta}+\delta_{i\alpha}l_j'(z_\beta)\partial_z\text{W}_{\alpha\beta}+\text{k}_y^2\text{W}_{\alpha\beta}\delta_{i\alpha}\delta_{j\beta}\right]\tilde{\text{G}}_{\alpha\beta}^{(e)}\\ &+\sum\nolimits_\alpha\nu_\alpha\text{W}_\alpha\tilde{\text{G}}_\alpha^{(e)}\Big\}=\text{W}(\boldsymbol{x}_s)/2.\end{aligned} \tag{21}$$

Here, $\tilde{\text{G}}_{\alpha\beta}^{(e)}$ are the discrete values of the electric potential spectra $\tilde{\text{G}}\left(x,\text{k}_y,z\right)$ in the subdomain Ω_e. Choosing the Lagrange basis polynomials as the weighting functions, i.e. $\{\text{W}_{pq}=l_p(x)l_q(z), \forall p, q\}$, Eq. (21) is changed into

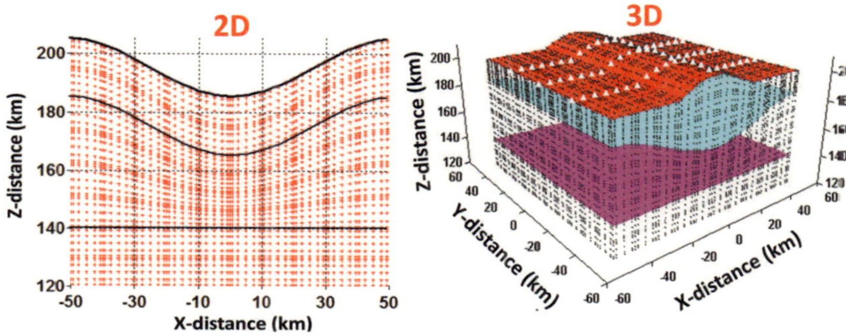

Figure 6.
2D and 3D Gaussian quadrature grids for numerical modelling: (a) a 2D model, 8 × 10 subdomains and 7 × 7 Gaussian abscissae in each subdomain and (b) a 3D model, 10 × 10 × 10 subdomains and 5 × 5 × 5 Gaussian abscissae in each subdomain.

$$\sum_e \Big\{ \sum_{\alpha,\beta,i,j} w_{\alpha\beta}\sigma_{\alpha\beta}\left[l_i'(x_\alpha)l_p'(x_\alpha)\delta_{j\beta}\delta_{q\beta} + l_j'(z_\beta)l_q'(z_\beta)\delta_{i\alpha}\delta_{p\alpha} + k_y^2\delta_{p\alpha}\delta_{q\beta}\delta_{i\alpha}\delta_{j\beta}\right]\tilde{G}_{\alpha\beta}^{(e)} + \sum_\gamma \nu_\gamma l_p(x_\gamma)l_q(z_\gamma)\tilde{G}_\gamma^{(e)} \Big\} = l_p(x_s)l_q(z_s)/2, \tag{22}$$

which can be rewritten in the matrix form for all points (p,q) and (i,j) in Ω:

$$\mathbf{M}\tilde{\mathbf{G}} = \mathbf{b}_s, \tag{23}$$

where **M** is a square symmetric matrix assembled by the coefficients of $\tilde{G}_{\alpha\beta}^{(e)}$ in Eq. (22). $\tilde{\mathbf{G}}$ and $\mathbf{b}_s$ are two vectors: the former consists of the discrete values of $\tilde{G}$ at all points of the Gaussian quadrature grid, and the latter is a zero vector except for the component of 1/2 at the electric current source located at $\boldsymbol{x}_s$. Eq. (22) shows that the matrices **M** and $\tilde{\mathbf{G}}$ depend on the wavenumber k_y and conductivity model $\sigma_{\alpha\beta}$. The wavenumber can be predicated by the minimum and maximum spaces of electrodes [17]. The linear equation system may be efficiently solved for multiple current sources $\mathbf{B} = \{\mathbf{b}_1,\mathbf{b}_2,\ldots,\mathbf{b}_N\}$ by the Cholesky decomposition [27]. After obtaining the spectra $\tilde{\mathbf{G}}$, the theoretical electric potentials are obtained by inverse cosine Fourier transform $\mathbf{U}(x,y,z) = I \cdot F_c^{-1}\left(\tilde{\mathbf{G}}\right)$.

From Eq. (22), one can find that the model conductivity $\sigma(x,z)$ is discretised by $\sigma_{\alpha\beta}$ at the Gaussian abscissae instead of finite elements. The subdomain integrals are calculated by the weights $w_{\alpha\beta}$ to the discrete integrands. The dense abscissae and variable $\sigma_{\alpha\beta}$ may describe the details of a complex geological model and generate high accurate solutions, so that sizes of the subdomains are not necessarily small but shaped to match the free-surface and subsurface interfaces. Apparently, accuracy of the numerical modelling depends on the number of the Gaussian abscissae in the subdomains. The more abscissae are applied, the more accurate results are yielded but cost more computer time. For efficient and accurate modelling, the minimum electrode space of electrode array may be chosen for the size of the subdomain, to which five Gaussian abscissae is applied to produce satisfactory solutions of numerical modelling. As examples, **Figure 7** shows the pseudo-sections of dipole–dipole, Schlumberger and gradient arrays for a layout of total 141 electrodes over an anticline model. These results show that similar pseudo-sections of apparent resistivity are obtained by using the three electrode arrays. From the above analysis, one

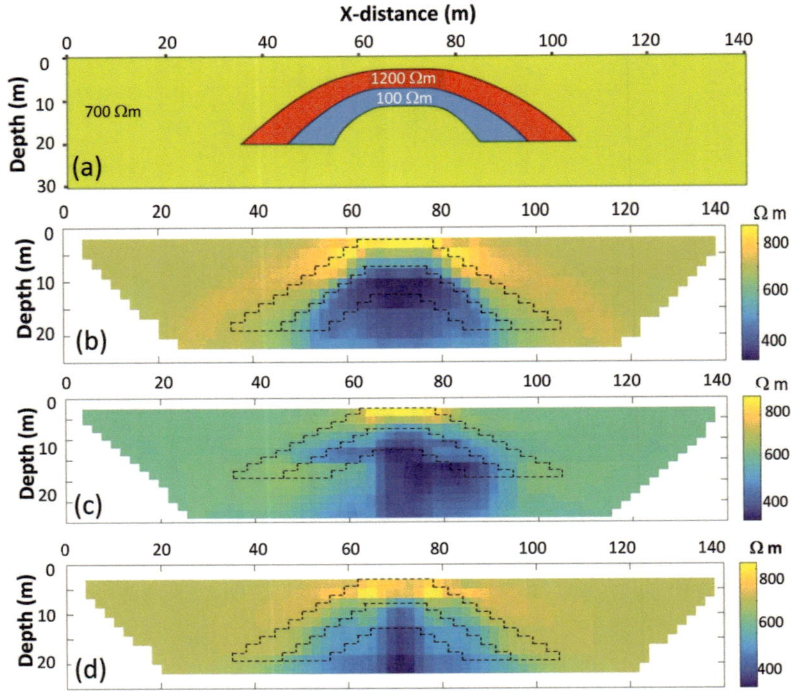

Figure 7.
Numerical modelling for (a) an anticline model and the apparent resistivity pseudo-sections of (b) dipole-dipole, (c) gradient and (d) Schlumberger arrays. The discrete anticline model is also given in the pseudo-sections.

can see that the finite-element approach does not calculate the high-order derivatives for the governing equation, but any finite-difference scheme does. Therefore, finite-element approach is often called a 'weak' solution of the governing equation against a 'strong' solution obtained by finite-difference method.

5. Tomographic inversion

Tomographic inversion is to reconstruct the geological model that offers synthetic data matching with observed data. Due to incompleteness and noise contamination of observed data, the model reconstruction is ill posed (multiple solutions). Therefore, tomographic inversion is often defined as an optimisation of data fittingness with regularisation of the model [28], e.g. a generalised objective function is applied:

$$\Phi(\mathbf{m}) = \left\|\mathbf{d}_{ob} - \mathbf{d}_{syn}(\mathbf{m})\right\|_{\mathbf{W}_d}^{l_p} + \lambda\|\mathbf{m}\|_{\mathbf{W}_m}^{l_p}, \tag{24}$$

where $\mathbf{d}_{ob}$ is a vector of observed data, which are either apparent resistivities or potential differences measured by different electrode arrays. $\mathbf{d}_{syn}(\mathbf{m})$ stands for synthetic data and is calculated by the finite-difference or finite-element method for a guessed geological model $\mathbf{m}$, which consists of discrete conductivities or resistivities. λ is a regularisation parameter that plays a trade-off role between the data fittingness (the first term in Eq. (24)) and model smoothness (the second term in Eq. (24)). $\|\cdot\|_W^{l_p}$ stands for the weighted l_p-norm with a weighting matrix $\boldsymbol{W}$, e.g. $\|\varepsilon\|_W^2 = \varepsilon^T \boldsymbol{W}\varepsilon$ and $\|\varepsilon\|_W^1 = \sum_i W_i|\varepsilon_i|$ are the weighted l_2-norm (generalised least

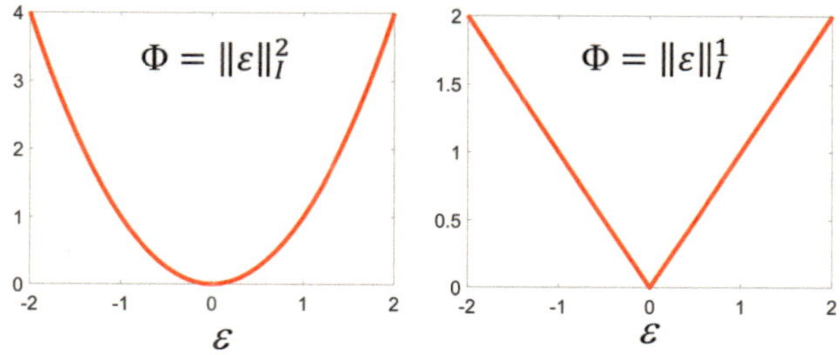

Figure 8.
An l_2-normed (left) and an l_1-normed (right) objective function. Here I is an unit matrix.

square) and the weighted absolute norm. **Figure 8** illustrates the difference between a l_2-normed and l_1-normed objective functions. $\boldsymbol{W}_d$ and $\boldsymbol{W}_m$ are weighting matrices to data and model, respectively. It is common to choose a combination of finite-difference operators for $\boldsymbol{W}_m$, e.g. $\boldsymbol{W}_m = \lambda_0 \boldsymbol{I} + \lambda_1 \mathbf{D}_x^T \mathbf{D}_x + \lambda_2 \mathbf{D}_y^T \mathbf{D}_y + \lambda_3 \mathbf{D}_z^T \mathbf{D}_z$ [29]. Here $\boldsymbol{I}$ is a unit matrix, and $\mathbf{D}_x$, $\mathbf{D}_y$ and $\mathbf{D}_z$ are the finite-difference operators in the x-, y- and z-directions. λ_k (k = 0,1,2,3) are constants and called extensional regularisation parameters used for searching for the smoothest model in three directions. Therefore, tomographic inversion becomes solving the following optimization problem:

$$\mathbf{m}^* = \min\{\Phi(\mathbf{m})\}. \quad (25)$$

To do so, a global or a local search may be applied to Eq. (25) [30], but the global search is extraordinarily computer time-consuming if $\mathbf{m}$ has a larger dimension [31]. Therefore, the local search of the standard conjugate gradient method is commonly applied for tomographic inversion. **Figure 9** gives a flowchart of the conjugate gradient algorithm and shows that the gradient $\nabla\Phi(\mathbf{m}_i)$ and the Hessian matrix $\mathbf{H}(\mathbf{m}_i)$ of the objective function are required. Applying the linearised approximation to the synthetic data, $\mathbf{d}_{syn}(\mathbf{m}_{i+1}) \approx \mathbf{d}_{syn}(\mathbf{m}_i) + (\frac{\partial \mathbf{d}_{syn}}{\partial \mathbf{m}})^T(\mathbf{m}_{i+1}\text{-}\mathbf{m}_i)$, and then substituting it for Eq. (24), one can obtain the gradient and the Hessian matrix [28]:

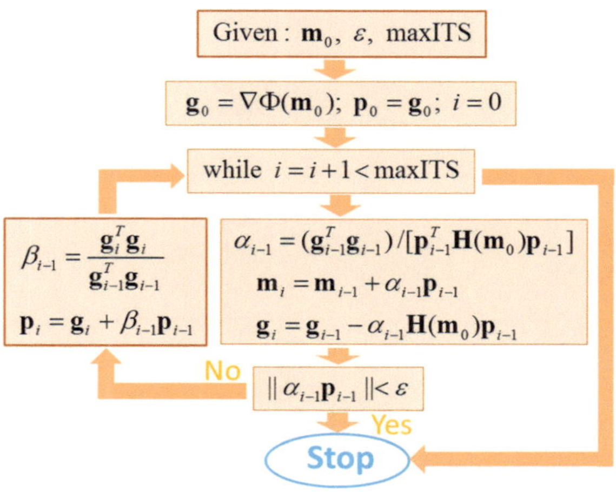

Figure 9.
Flowchart of a standard conjugate gradient algorithm for tomographic inversion. Here $\mathbf{m}_0$ is an initial model, ε is a small value, maxITS is the maximum iterations, and $\mathbf{H}(\mathbf{m}_i)$ is the Hessian matrix of object function $\Phi(\mathbf{m}_i)$.

$$\nabla \Phi(\mathbf{m}_i) = \left(\frac{\partial \mathbf{d}}{\partial \mathbf{m}_i}\right)^T \mathbf{W}_d \left[\mathbf{d}_{ob} - \mathbf{d}_{syn}(\mathbf{m}_i)\right] + \lambda \mathbf{W}_{\mathrm{m}}(\mathbf{m}_i - \mathbf{m}_0), \tag{26}$$

$$\mathbf{H}(\mathbf{m}_i) = \left(\frac{\partial \mathbf{d}}{\partial \mathbf{m}_i}\right)^T \mathbf{W}_d \frac{\partial \mathbf{d}}{\partial \mathbf{m}_i} + \lambda \mathbf{W}_{\mathrm{m}}, \tag{27}$$

for the l_2-normed objective function.

Note that the gradient $\nabla\Phi(\mathbf{m}_i)$ of the l_1-normed objective function has multiple values at zero misfit $\Phi(\mathbf{m}_i) = 0$ (see **Figure 8**), but for any non-zero misfit $\Phi(\mathbf{m}_i) \geq \varepsilon_0$ (ε_0 is a very small value), the computations of the gradient $\nabla\Phi(\mathbf{m}_i)$ and the Hessian matrix $\mathbf{H}(\mathbf{m}_i)$ become simple, and they are similar to Eqs. (26) and (27) except for the weighting matrices $\mathbf{W}_d$ and $\mathbf{W}_{\mathrm{m}}$. Therefore, as $\Phi(\mathbf{m}_i) < \varepsilon_0$, the l_1-normed gradient $\nabla\Phi(\mathbf{m}_i)$ and the Hessian matrix $\mathbf{H}(\mathbf{m}_i)$ may be replaced with Eqs. (26) and (27), and as $\Phi(\mathbf{m}_i) \geq \varepsilon_0$ the l_1-normed gradient and the Hessian matrix are directly applied to the standard conjugate gradient algorithm. Therefore, the l_1-normed inversion and the l_2-normed inversion are implemented with an almost same algorithm. However, their inversion results may be quite different in the case that the observed data have outliers. Many synthetic and practical experiments have shown that the l_1-normed inversion is less sensitive to the outliers [13]. If the data have no outliers or high qualities, two inversions converge [11].

Figure 10 shows synthetic experiments for imaging the anticline structure shown in **Figure 7a** with dipole–dipole, Schlumberger and gradient arrays. The l_2-normed inversions were conducted with the apparent resistivities shown in **Figure 7b–d**, which were computed with the GQG software. The commercial software RES2DINV was applied to the inversions. From these results, one can see that the dipole-dipole and gradient arrays yield competitive images of the anticline structure. **Figure 11** gives real applications of surface and cross-hole ERT conducted at a rural site in Australia, where there are many existing drill boreholes which can be used for examinations of both surface and cross-hole ERT experiments. From these boreholes and logging data, the subsurface rocks and main structures were well known (shown in **Figure 11b**). The main objective of this work was to use the borehole geological and logging information and exam the imaging capabilities of surface and cross-hole ERT in this area, particularly for mapping the base of alluvial overburden and the base of pisolite, as well as predicting clay contamination within pisolite. Surface data acquisition with dipole-dipole and Schlumberger arrays was

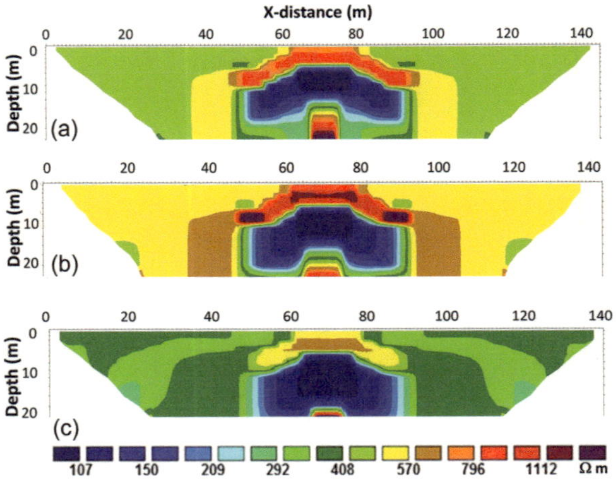

Figure 10.
Tomographic inversion of (a) dipole-dipole, (b) Schlumberger and (c) gradient arrays for the anticline model shown in ***Figure 7****.*

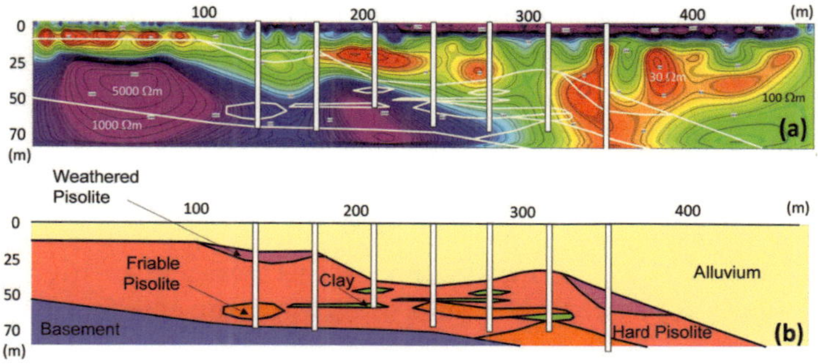

Figure 11.
Applications of surface and cross-hole electrical resistivity tomography for mapping base of alluvial overburden: (a) integrated resistivity images from surface and cross-hole ERT and (b) geological section from borehole rock samples and logging data.

conducted along a 750 m line, and two pairs of cross-hole ERT on the same line were also carried out for details of the geological structure between the boreholes. **Figure 11a** shows the integrated cross-hole ERT results with the surface ERT imaging. It gives the resistivity structure along the line from the surface and cross-hole ERT. **Figure 11b** is the geological section from the existing boreholes and logging data along the same line. Comparing the resistivity imaging results with the geological section, one can see that the integrated ERT results well map the base of alluvial overburden and the base of pisolite. The clay contamination within pisolite is also shown in cross-hole ERT images from the horizontal distance 200–300 m.

6. Conclusions

ERT is a useful near-surface imaging technique, which mainly include data acquisition, numerical modelling and tomographic inversion. For surface data acquisition, dipole-dipole, Schlumberger and gradient arrays are applicable for high-resolution image; particularly, the full-range gradient array may complement to gradient array for better data coverage and completeness of data information. For cross-hole data acquisition, pole-pole (A-M), bipole-pole (AM-N) or pole-bipole (N-AM) and bipole-bipole (AM-BN) can be employed, and the geometry factor and numerical modelling may be applied for designing efficient and effective arrays and exam the imaging capability of ERT for specified targets.

GQG approach is a new version of finite-element modelling. It uses Gaussian abscissae to discrete the model domain and Gaussian weights to compute the volume integral. Therefore, it is much easier to match arbitrary free-surface topography and subsurface interface and computation of the subdomain integrals. It does not require a small size of element and complex element mesh generator. The accuracy of modelling depends on the number of Gaussian abscissae in subdomains. The more abscissae are employed in subdomains, the more accurate modelling result is generated but costs more computer time.

Tomographic inversion is generally implemented by a standard conjugate gradient algorithm, which requires to compute the gradient and the Hessian matrix of an objective function. Two types of objective functions can be applied. One is the l_1-norm and the other is the l_2-norm. The former is less sensitive to an outlier in data but hardly computes the gradient at zero misfit. The latter has no problem to compute the gradient and the Hessian matrix, and it is much easier to implement a standard conjugate gradient algorithm. With high-quality data, the two inversions converge. The field experiments show that surface and cross-hole ERT can be applied to map the base of alluvial overburden and the base of pisolite, as well as the clay contamination within pisolite.

Acknowledgements

The author thanks the Abu Dhabi Education Council for the Award for Research Excellence (AARE17-273) to financially support this work and greatly appreciates the ROBE company for sharing the results of ERT imaging experiments.

Author details

Bing Zhou
Department of Earth Science, Khalifa University of Science and Technology, Abu Dhabi, UAE

*Address all correspondence to: bing.zhou@ku.ac.ae

References

[1] Smith NC, Vozoff K. Two-dimensional DC resistivity inversion for dipole–dipole data. IEEE Transactions on Geoscience and Remote Sensing. 1984;**GE-22**:21-28

[2] Sasaki Y. Resolution of resistivity tomography inferred from numerical simulation. Geophysical Prospecting. 1992;**40**:453-464

[3] Dahlin T. 2D resistivity surveying for environmental and engineering applications. First Break. 1996;**14**:275-283

[4] Zhe J, Greenhalgh SA, Marescot L. Multi-channel, full waveform and flexible electrode combination resistivity imaging system. Geophysics. 2007;**72**:F57-F64

[5] LaBrecque D, Miletto M, Daily W, Ramirez A, Owen E. The effects of noise on "Occam" inversion of resistivity tomography data. Geophysics. 1996;**61**:538-548

[6] Loke MH, Barker RD. Rapid least-squares inversion of apparent resistivity pseudosections by a quasi-Newton Method. Geophysical Prospecting. 1996;**44**:131-152

[7] Mauriello P, Monna D, Patella D. 3-D geoelectric tomography and archaeological applications. Geophysical Prospecting. 1998;**46**:543-570

[8] Mauriello P, Patella D. Resistivity anomaly imaging by probability tomography. Geophysical Prospecting. 1999;**47**:411-429

[9] Zhou B, Greenhalgh SA. Explicit expressions and numerical calculations for the Fréchet and second derivatives in 2.5D Helmholtz equation inversion. Geophysical Prospecting. 1999;**47**:443-468

[10] Loke MH, Chambers JE, Rucker DF, Kuras O, Wilkinson PB. Recent development in the direct-current geoelectrical imaging method. Journal of Applied Geophysics. 2013;**95**:135-156

[11] Dahlin T, Zhou B. A numerical comparison of 2D resistivity imaging with 10 electrode arrays. Geophysical Prospecting. 2004;**52**:379-398

[12] Dahlin T, Zhou B. Multiple-gradient array measurements for multichannel 2D resistivity imaging. Near Surface Geophysics. 2006;**4**:113-123

[13] Zhou B, Dahlin T. Properties and effects of measurement errors on 2D resistivity imaging surveying. Near Surface Geophysics. 2003:105-117

[14] Zhou B, Greenhalgh SA. Cross-hole resistivity tomography using different electrode configurations. Geophysical Prospecting. 2000;**48**:887-912

[15] Zhou B, Greenhalgh SA. Rapid 2-D/3-D crosshole resistivity imaging using the analytic sensitivity function. Geophysics. 2002;**67**:755-765

[16] Mufti IR. Finite difference resistivity modelling for arbitrary shaped two-dimensional structures. Geophysics. 1976;**41**:62-78

[17] Dey A, Morrison HF. Resistivity modelling for arbitary shaped two-dimensional structures. Geophysical Prospecting. 1979;**27**:106-136

[18] Dey A, Morrison HF. Resistivity modelling for arbitrary shaped three-dimensional structures. Geophysics. 1979;**44**:753-780

[19] Mundry E. Geoelectrical model calculations for two-dimensional resistivity distributions. Geophysical Prospecting. 1984;**32**:124-131

[20] Spitzer K. A 3-D finite difference algorithm for DC resistivity modelling using conjugate gradient methods. Geophysical Journal International. 1995; **123**:903-914

[21] Zienkiewicz OC. The Finite Element Method in Engineering Science. London, New York: McGraw-Hill Book Co, 1971

[22] Coggon JH. Electromagnetic and electrical modelling by the finite element method. Geophysics. 1971;**36**: 132-155

[23] Fox RC, Hohmann GW, Killpact TJ, Rijo L. Topographic effects in resistivity and induced polarization surveys. Geophysics. 1980;**45**:75-93

[24] Pridmore D, Hohmann GW, Ward SH, Sill WR. An investigation of the finite element method for electrical and electromagnetic modelling data in three dimensions. Geophysics. 1981;**46**: 1009-1024

[25] Queralt P, Pous P, Marcuello A. 2D resistivity modelling: An approach to arrays parallel to the strike direction. Geophysics. 1991;**56**:941-950

[26] Zhou B, Greenhalgh SA. Finite element three-dimensional direct current resistivity modelling: Accuracy and efficiency considerations. Geophysical Journal International. 2001; **145**:676-688

[27] Press WH, Teukolsky SA, Vetterling WT, Flannery BP. Numerical Recipes in C: The Art of Scientific Computing. Second ed. England: Cambridge University, EPress; 1992

[28] Greenhalgh SA, Zhou B, Green A. Solutions, algorithms and inter-relations for local minimization search geophysical inversion. Journal of Geophysics and Engineering. 2006;**3**: 101-113

[29] Ellis RG, Oldenburg DW. The pole-pole 3-D-resistivity inverse problem: A conjugate-gradient approach. Geophysical Journal International. 1994; **119**:187-194

[30] Kirsch A. An Introduction to the Mathematical Theory of Inverse Problems. Basel: Springer; 1996

[31] Schwarzbach C, Ralph-Uwe B, Klaus S. Two-dimensional inversion of direct current resistivity data using a parallel, multi-objective genetic algorithm. Geophysical Journal International. 2005;**162**:685-695